DESIGNING ONLINE CONTINUING PROFESSIONAL DEVELOPMENT FOR STEM TEACHER EDUCATION

Bridging the realms of educational technology and teacher education, this book offers instructional design models, strategies, and tools for online delivery of continuing professional development (CPD) for science, technology, engineering, and mathematics (STEM) teachers. Acknowledging that successful STEM education demands strategies grounded in specific instructional design frameworks, this complex integration of pedagogical strategies with teaching and facilitation practices creates unique challenges for developing effective online learning environments.

In this volume, editors Heejung An and Woonhee Sung have invited STEM teacher educators to share feasible models, strategies, and research-proven practices to enhance instructors' knowledge and practices and enable them to design, develop, and implement their own online CPD. To that end, chapters contextualize online STEM education course designs and teaching trends across degree-based continuing education, professional development, and teacher professional learning. The chapters focus on the most essential pedagogical components of STEM education CPD, including hands-on, real-world-based projects, learner practice and reflection, interaction with peers and instructors, as well as how to effectively design and deliver these elements. The practical applications in the various chapters demonstrate how online programs can use various technological tools to educate STEM teachers in online settings with a variety of STEM materials.

The book also provides concrete lesson plans that online faculty and CPD providers can apply in any teaching setting to strengthen STEM teacher educators' capabilities. Ideal for instructors of CPD who design and teach STEM-related educational disciplines, this unique collection responds to the need to bring online and blended learning to the STEM education community while maintaining quality and standards-based content acquisition.

Heejung An is a Professor and Chair in the Department of Educational Leadership and Professional Studies in the College of Education at William Paterson University of New Jersey.

Woonhee Sung is an Assistant Professor of instructional technology in the School of Education at The University of Texas at Tyler.

Designing Online Continuing Professional Development for STEM Teacher Education: Models, Strategies, and Practices is a timely and essential contribution to the field. The editors and contributors offer research-informed models, strategies, and real-world examples that not only respond to the challenges of online learning but also reimagine what professional growth for STEM educators can look like in inclusive, digitally connected contexts. This volume bridges theory and practice with clarity and purpose, equipping teacher educators, practitioners, and instructional designers with concrete tools for fostering innovation, equity, and collaboration. It is an indispensable resource for anyone committed to advancing meaningful and accessible professional learning in STEM education.

— **Matthew Farber**, EdD, Associate Professor of Educational Technology, University of Northern Colorado

Designing Online Continuing Professional Development for STEM Teacher Education is a practical bridge between research rigor and classroom reality. It integrates learning theories, instructional design, technology integration, and STEM pedagogy into a clear, step-by-step guide for online professional development. Grounded in constructionist and design-based learning, it centers on accessibility, culturally responsive practice, and meaningful assessment. This book is essential reading for faculty, instructional designers, and district leaders building scalable and equitable online professional development in teacher education.

— **Nari Kim**, PhD, Professor, School of Education and Human Services, University of Wisconsin–Oshkosh

Designing Online Continuing Professional Development for STEM Teacher Education is a timely and useful book for anyone seeking to enhance STEM education in contemporary schools, which are increasingly becoming more diverse and interested in new technologies. Editors Heejung An and Woonhee Sung have brought together leading researchers and experts to share research-backed models and practical methods, such as flipped learning, R2D2, TMSR, place-based learning, and AI-infused approaches, that make online professional development easy to access and beneficial. What makes this book different is its strong dedication to diversity, equity, inclusion, and accessibility. This means that every administrator and teacher may find resources and motivation to help all of their children. This book provides you with the vision and real-life examples you need to create online professional learning that truly transforms your school teaching. Curriculum designers and school administrators will also find useful information here.

— **Marc A. Medley**, Principal, Paterson Public Schools; Book Talk Radio Host, 88.7 FM and gobrave.org (Brave New Radio)

This timely book addresses today's rapid advancement of technology and offers practical and plentiful resources for the professional development of STEM educators who are struggling to keep apace. The authors understand that online environments require attention to thoughtful instructional design that ensures meaningful teaching and learning. They offer research informed methods, models, and strategies for Continuous Professional Development of STEM educators and administrators. This book is an excellent resource for not only STEM educators but for anyone interested in staying informed on best practices in STEM education and online professional development.

— **Geraldine Mongillo**, PhD, Professor Emerita, Department of Educational Leadership and Professional Studies, William Paterson University

Designing Online Continuing Professional Development for STEM Teacher Education is an indispensable resource that bridges research and practice to support educators, leaders, and instructional designers. With research-informed models, real-world case studies, and a strong commitment to diversity, equity, inclusion, and accessibility, this book offers practical strategies for designing meaningful and impactful online professional learning. It is a must-read for anyone dedicated to advancing STEM education through innovative and inclusive professional development.

— **Meina Zhu**, PhD., Associate Professor, Wayne State University

DESIGNING ONLINE CONTINUING PROFESSIONAL DEVELOPMENT FOR STEM TEACHER EDUCATION

Models, Strategies, and Practices

Edited by Heejung An and Woonhee Sung

Routledge
Taylor & Francis Group
NEW YORK AND LONDON

Designed cover image: © Getty Images

First published 2026
by Routledge
605 Third Avenue, New York, NY 10158

and by Routledge
4 Park Square, Milton Park, Abingdon, Oxon, OX14 4RN

Routledge is an imprint of the Taylor & Francis Group, an informa business

ISBN: 9781032869919 (hbk)
ISBN: 9781032869902 (pbk)
ISBN: 9781003530343 (ebk)

DOI: 10.4324/9781003530343

Typeset in Sabon
by KnowledgeWorks Global Ltd.

CONTENTS

FOREWORD

The title of this book—*Designing Online Continuing Professional Development for STEM Teacher Education*—presents a challenging paradox. On the one hand, professional development, bolstered by the COVID-19 pandemic, is increasingly delivered through online tools. On the other hand, professional development for science, technology, engineering, and mathematics (STEM) education benefits greatly from direct, hands-on experiences. And what about the role of AI in professional development? Indeed, since the 2022 release of ChatGPT 3.5, artificial intelligence has gained a significant presence in professional development, and this trend is likely to continue. So how can all this be reconciled? Can there be effective online professional development (sometimes AI-enhanced) for a topic like STEM education? Essentially, this book is designed to address these questions.

The book's lead editors, Heejung An and Woonhee Sung, are wise to tackle these issues upfront in an introductory chapter (authored by An and Chang), called "Setting the Stage: Supporting STEM Teachers Through Online Continuing Professional Development in a Diverse and Digitally Connected World." This chapter establishes key themes that are carried through in subsequent chapters: online tools allow for greater global connections; virtual environments provide for participation of diverse populations (subject to meaningful access to online resources); the power of citizen science and crowdsourcing are well-suited for online environments. The roles of diversity and equity are consistently addressed by other chapter authors.

Notably, An and Sung divide their book into two sections: the first, which deals with the effective design of professional development, and the second, which addresses issues related to equity in online professional development. They present a network of well-regarded chapter authors, who address prominent issues in contemporary education models, ranging from TPACK to Authentic Learning, from Attack to Flipped Learning. Overall, the issues they address are highly relevant as educators worldwide struggle with the future of professional development, particularly in an AI-enhanced world.

Interestingly, as this foreword was being prepared, this author attended the National Technology Leadership Summit (NTLS; ntls.info) in Washington, DC. At that meeting, important models that blended professional development for STEM, online learning, and even

AI were recurring areas of discourse. For example, the BranchED organization developed a new tool that combined a virtual avatar "coach" to guide a student teacher on how to address a student with certain misconceptions in math knowledge. After the counseling session, the online tool was then used to present the student teacher's revised interaction to correct her student's initial misconception. BranchED plans to increase the role of AI-based scripts to enhance this work.

An entire strand at NTLS—Reciprocal Innovation through NTLS Innovation Centers—was devoted to international STEM activities, utilizing online and hands-on tools and interactions. The strand highlighted years-long work, led by a team from the University of Virginia, that combined STEM, makerspaces, citizenship science, and even AI. The work has increasingly included international partners, most recently a delegation from Uganda. And this is where the work of An and Sung, and their colleagues, is critical. For complex, collaborative *international* projects to succeed, they will inevitably include some degree of online training and professional development. In STEM-based projects, there are almost always moments when a lead teacher/trainer demonstrates the use of certain tools essential to the project's success. And, unless this training takes place in a single location, the use of online tools is essential, especially for an international project. Of course, exchange arrangements where students and teachers can visit each other's spaces can be considered; most interactions—including professional development—will take place through online tools. Over time, as discussed above, it's likely that AI will play an increasingly significant role in both. For example, could AI-based real-time, on-the-fly solutions be constructed to address the inevitable concerns and problems that will arise with STEM activities?

Interestingly, the use of the term *reciprocal* in the project described above has led to a team discussion on Robert Taylor's *The Computer in the school: Tutor, tool, tutee.* Both reciprocal innovation and reciprocal learning suggest that the learning opportunities for project members, ranging from students to teachers and trainers, are negotiable and fluid. At various times, roles and activities are brokered, from teacher to learner and vice versa. Likewise, in An's preface, she describes how Taylor's 3T model led her to realize "more fully that technology in my courses could act not only as a tool but also as a tutor that guided learning and even as a tutee shaped by students' input." Ideally, a blended AI online learning space allows learners and teachers to seemingly negotiate their roles in a seamless manner.

Recently, a problem arose in student coding for the "reciprocal innovation" project, described above, where students were tasked with using block coding to control certain optical equipment. One team's block code would not execute the intended function. Then, another team member took the code, translated it into the programming language LISP, and fed it into ChapGPT. The GenAI tool then guided him through the debugging process, which led to a solution, enabling the successful execution of the optical instrumentation project. Note that the "reciprocal innovation" project is an international one, across thousands of miles and multiple time zones. As the book's introductory chapter notes, much of contemporary professional development and training will occur within a "… diverse and digitally connected world." The ability to apply AI-based real-time solutions to problems encountered in STEM projects can complement formal and prescriptive professional development activities. And, as the GenAi tool acted as a guide, not providing the direct solution, but enabling the team to work through the process, we're reminded of Taylor's 3T model and interactions with AI tools that highlight the *tutor* and *tutee.*

Of course, it's too much to ask that the *Designing Online Continuing Professional Development for STEM Teacher Education* book resolve all these issues. Yet, professors Heejung An and Woonhee Sung, along with their team, provide insightful discussions that highlight these matters and set a path for informed discourse and future investigations.

Michael Searson, PhD
Director of Program Development, Association for the Advancement of Computers in Education (AACE)

Reference

Taylor, R. P. (Ed.). (1980). *The computer in school: Tutor, tool, tutee*. Teachers College Press.

PREFACE

I was first introduced to online teaching and learning in 1999, when I took the course MSTU 4022: *Telecommunications and Distance Learning* with the late Professor Robert P. Taylor at Teachers College, Columbia University, during my doctoral program. Professor Taylor, one of the international digital pioneers, developed the conceptual framework for the potential functions of the computer in educational settings as tutors, tools, and tutees (the 3T model), in his 1980 book *The Computer in the School*. In this work, he explained how the 3T model could provide a framework for understanding the ways technology might support technology-enriched learning. Through his course, I also learned how the Internet could be leveraged for collaborative exchanges between the United States and Africa, as well as the broader educational implications of what was then referred to as distance learning. At that time, however, I perceived technology primarily as a tool for facilitating instruction, rather than as a broader framework for transforming teaching and learning. Even so, being introduced to Taylor's 3T model laid the groundwork for my later recognition that computers can serve multiple roles in shaping technology-enhanced learning environments.

After completing this course, I was fortunate to be invited to work on the "Distance Learning Project" at Teachers College. Although my involvement was minimal, this experience marked the first time I started to understand the potential of online learning and how universities might effectively leverage online instruction. I later enrolled in several online courses, primarily offered by the Department of Human Development. At that time, the design of these courses was quite passive: Professors recorded lectures that students listened to via audio files while following along with text- or audio-based lecture notes, or students simply read assigned articles and posted weekly responses on a discussion board.

Together, these experiences sparked my interest in studying how online environments could be made more interactive and effective and led to my dissertation study, *Designing an Effective Web-Based Coding Environment for Novice Learners: The Effects of Example Representations*, which focused on helping learners build mental models of how coding works while examining the effects of example representations and immediate feedback. Building on this line of research, I also began exploring how universities were offering online courses. This exploration opened opportunities for an adjunct faculty position at

Grand Canyon University and Stony Brook University, where I taught online courses for in-service teachers across several U.S. states.

When I joined William Paterson University of New Jersey in 2005, online learning was not yet a prominent part of the institutional landscape, and my engagement with online teaching had paused. However, years later, the COVID-19 pandemic reignited my interest and commitment as I realized that my previous research and teaching experiences were coming together with my work in STEM education and educational technologies. Because the content I teach typically involves inductive, inquiry-based, and project-oriented learning, simply reading articles and posting responses was insufficient to replicate the richness of in-person learning. As a result, I had to rethink and redesign the teaching process to ensure that online environments could still support meaningful learning experiences and outcomes. In doing so, I found myself returning to Professor Taylor's 3T model, realizing more fully that technology in my courses could act not only as a tool but also as a tutor that guided learning and even as a tutee shaped by students' input. In addition, I became more intentional about preparing teachers to be able to apply what they have learned in online environments to their classrooms. Through these efforts, I explored various instructional design models and evidence-based approaches to create effective online STEM-oriented courses for teachers. In the process, I discovered that many other faculty members and instructional designers were facing similar challenges.

Over the years, I have witnessed both the successes and the shortcomings of online learning, continually reflecting on which instructional design models and strategies, along with emerging technologies, might shape online continuing professional development (CPD) in the coming decades. In many ways, Professor Taylor's early vision of computers serving as tutors, tools, and tutees has become a reality in today's online STEM CPD.

This realization, and my experiences in the field over the years, inspired me to envision this edited volume as a resource for others navigating the complexities of designing and facilitating online CPD for STEM educators. The 13 chapters in this edited volume represent the culmination of research, practice, and reflection from authors focusing on online STEM CPD. It presents research-informed models and strategies for designing and facilitating effective online CPD for STEM educators, with several chapters explicitly foregrounding the principles of diversity, equity, inclusion, and accessibility.

In Chapter 1, which serves as the foundation for the book, Heejung An and Ting-Han Chang provide an overview of the importance and scope of STEM education, along with the growing need for online continuing professional development to advance STEM teachers' practices and knowledge. The chapter also highlights the significance of addressing diversity, equity, inclusion, and accessibility when designing online professional development programs.

Section I. Designing Effective Online Continuing Professional Development for STEM Educators: Models, Strategies, and Case Examples

The chapters within this section describe instructional design models and explore design principles and pedagogical approaches appropriate for STEM-focused online CPD for teachers. Additionally, this section features real-world CPD projects and modules across various STEM contexts, highlighting specific instructional design models and strategies used to illustrate practical applications and implementations.

In Chapter 2, Dana Smerda-Mason introduces the 5Ps (the purpose, place, platform, population, and profit) Model as a helpful framework for designing efficient and impactful virtual communities for STEAM educators.

In Chapter 3, Woonhee Sung and Heejung An present a design-based methodology for online CPD that incorporates the arts into STEM disciplines via the TMSR (Thinkering, Making, Sharing, Reflection) Model and the 4-Ons framework (hands-on, minds-on, hearts-on, social-on). This approach aims to promote collaboration, creativity, and pedagogical innovation within STEAM education.

In Chapter 4, Gloria Campbell-Whatley and Kimberly Bunch-Crump introduce the ATTACK (Application, Teaching Strategy, Trial/Try-Out, Assessment, Cognitive Reflection, and Keep/Retain/Generalize) Model for planning and delivering literature-integrated STEM instruction, with the goal of enhancing content accessibility, cultural relevance, and student engagement through the integration of real-world narratives into scientific instruction.

In Chapter 5, Bekir Yıldırım presents the online flipped STEM Teacher Training Institute model, which integrates flipped learning with eight developmental stages to support high-quality professional development, illustrated through an instructional example and recommendations.

In Chapter 6, Heejung An describes how the R2D2 (Read, Reflect, Display, and Do) Model can be used to design an online CPD that supports diverse learning preferences and actively engages learners in the process. It provides detailed explanations, instructional examples, and resources for each quadrant of the R2D2 Model.

In Chapter 7, Shing F. Fung, Chuck Higgins, Shirsh Lata Soni, Hilarie Davis, and Carolyn Ng introduce video-based, self-paced learning modules developed for Radio JOVE, one of the oldest NASA-funded citizen science programs, designed to prepare participants to build and operate inexpensive radio telescopes, make radio observations, and analyze collected data.

In Chapter 8, Heejung An and Sue-Anne Alonso describe exemplary practices illustrating how the TPACK Model can be applied to design science instruction that integrates artificial intelligence tools to support inquiry-based learning.

In Chapter 9, Tiffany A. Roman and Brendan E. Callahan explore an online professional development program that supports secondary science teachers in integrating sustainability concepts with artificial intelligence tools, aligned to Next Generation Science Standards and United Nations Sustainable Development Goals.

In Chapter 10, Thomas Chandler, Joshua DeVincenzo, and Shuyang Huang describe how the Model of Authentic Learning can be applied to the design of online professional development focused on hazard-resilient housing.

In Chapter 11, Woonhee Sung introduces the Extended TPACK (E-TPACK) framework, which addresses the dual-layer complexity of online STEM professional development. The framework underscores that CPD providers must design and facilitate effective virtual learning experiences using digital tools and online pedagogies, while also preparing K-12 STEM teachers to integrate content knowledge, technologies, and pedagogical strategies in their own classroom practice.

Section II. Support Materials to Design Equitable Online Continuing Professional Development for STEM Instruction

This section provides practical tools and resources, including ready-to-use rubrics and curated open educational resources (OERs), that instructors and instructional designers can leverage when designing online STEM CPD.

In Chapter 12, Siri Anderson and Barbara Billington stress the importance of STEM teachers advocating for gender-inclusive methods in their classrooms and provide evidence-based rubrics designed to help teachers be more effective in teaching female students, while also having a positive impact on students' STEM identities.

In Chapter 13, Yujin Park, Jewoong Moon, and Hunhui Na explore how OERs and Professional Learning Networks support STEM teachers' professional growth. They also highlight five key categories of OER: course materials, instructional strategies, lesson plans, professional development, and tools/tips.

Target Audience

The book was envisioned to primarily support the following three audiences: (a) faculty and graduate students in teacher education and education leadership programs; (b) K-12 STEM teachers and educational specialists such as curriculum developers, content-area specialists, school department heads, curriculum supervisors, and school administrators; and (c) instructional designers and instructors of professional development.

Although each audience may approach this book from a different perspective, the models and strategies are designed to promote collaboration across roles, bridging research and practice, linking classroom realities with system-level leadership, and aligning instructional design with teacher and student needs. We encourage readers to not just passively read the book, but to actively use the models and strategies in their own contexts, contemplate the results, and add to the ongoing professional conversation about how to best provide online continuing professional development for STEM teachers. Through this engagement, readers can join a larger community of professionals who are dedicated to promoting innovation and improving STEM education around the world. My hope is that educators will continue to harness this potential in ways that bring lasting value to both STEM classrooms and professional practice.

Heejung An
William Paterson University of New Jersey

Reference

Taylor, R. P. (Ed.). (1980). *The computer in school: Tutor, tool, tutee*. Teachers College Press.

AI DECLARATION

Some images in this book were generated using artificial intelligence technology. All ideas, concepts, research, analysis, and written content are entirely the original work of the author(s). The use of AI-generated imagery serves solely as visual support for the educational content and does not represent or influence the intellectual contributions, methodologies, or conclusions presented in this work.

ACKNOWLEDGMENTS

We would first like to express our heartfelt appreciation to our families, whose unwavering love, unconditional support, and trust have sustained our inner strength and commitment to this work. We are also deeply grateful to our students—in-service teachers, teacher leaders, and supervisors—who inspired the vision behind and creation of this book. Their dedication and passion for teaching and learning have been constant sources of insight, helping us grow as educators and scholars. In addition, we extend our sincere thanks to Routledge and the anonymous reviewers of the book proposal, whose thoughtful feedback was carefully considered and incorporated to strengthen this work.

Lastly, we share a common bond as graduates of the Instructional Technology & Media program in the Department of Mathematics, Science, and Technology at Teachers College, Columbia University. Looking back, we recognize how far-sighted and innovative the program was. The professors were pioneers in the field, and the courses we took not only provided authentic and rigorous theoretical foundations for our future careers and academic endeavors but also fostered a lifelong commitment to growth and the cultivation of new knowledge and skills. We remain grateful to Teachers College and to the professors who guided us.

The first editor, Heejung, would also like to express her appreciation to her two dogs, Max and the late Sammy, who slept by her desk through many long nights. Their quiet presence brought comfort and companionship throughout the writing and editing process.

Heejung An, William Paterson University of New Jersey
Woonhee Sung, The University of Texas at Tyler

CONTRIBUTING AUTHORS

Sue-Anne Alonso, MEd, is a Science Teacher at Wayne Valley High School, New Jersey, with an additional certification in ESL Curriculum and Learning, nurturing both content-area instruction and language learning. A graduate of Montclair State University's Woodrow Wilson Teaching Fellowship, she has presented best practices in science and language teaching at local, state, and international conferences. Her interest in transformative advocacy in action led to the development of a multilingual newscast and the creation of events designed to draw families into STEM education. She has developed curriculum at the middle and high school level, including extracurricular units on biomimicry and microscopy. As a Fulbright-Hays Groups Project Abroad scholar, she has taught, studied, and explored the South Korean educational system.

Heejung An, EdD, is a Professor and Chair of the Department of Educational Leadership and Professional Studies in the College of Education at William Paterson University of New Jersey. She teaches graduate courses for in-service teachers and has led numerous teacher professional development initiatives through grants such as the Inclusive Arts Integration Grant Program (2017–2023), funded by the Geraldine R. Dodge Foundation, and the Fulbright-Hays Group Projects Abroad Program (2013–2015), funded by the U.S. Department of Education. She also founded the annual William Paterson University's Educational Technology Virtual Conference in 2010, providing extensive professional learning for New Jersey educators (https://www.wpuedtechconference.com/).

Siri Anderson, EdD, has been in online teacher-licensure and graduate teacher education for more than 20 years. She has developed online graduate certificates for teachers in Computational Thinking and Coding, STEM, Technology Education, Literacy, Native American Pedagogy, and Curriculum and Instruction. Serving as the Co-Principal Investigator for NSF #1513060 Sci Girls Strategies: Gender Equitable Teaching Practices in Career and Technical Education Pathways for High School Girls for five years. She has been an advocate for single-gender learning experiences for decades and is proud to be part of the 7% of OUPV (Operator of Uninspected Passenger Vessel)-licensed captains by the U.S. Coast Guard who are female. She offers lessons and charter sailing on Lake Superior.

Barbara Billington, PhD, is an Education Program Specialist in the Center for Educational Innovation (CEI) at the University of Minnesota, where she provides programming and consulting in various STEM fields. With expertise in designing curriculum and learning environments that aspire to reach and teach all students, Dr. Billington supports educators of all levels. She has taught STEM students ages five to adult and relishes educating instructors in higher education. Prior to working in CEI, she taught preservice science teachers for 16 years. Her work with Twin Cities PBS on their SciGirls curriculum has afforded her the opportunity to provide workshops across the country.

Kimberly Bunch-Crump, PhD, is an Assistant Professor in the Department of Special Education and Child Development at the University of North Carolina at Charlotte. Her scholarship centers on personnel preparation, multi-tiered systems of support, applied behavior analysis, and culturally responsive practices in K–12 and higher education settings. She supports the preparation of undergraduate, graduate, and doctoral students for education and behavior specialist careers. Informed by her background as a staff development coach in K-12 and behavioral health settings, she develops and leads grant-funded initiatives equipping professionals to implement evidence-based, culturally responsive practices across diverse learning environments.

Brendan Callahan, PhD, is a Professor of Biology Education at Kennesaw State University. He earned his PhD in Curriculum and Instruction from the University of South Florida, where he investigated epistemological constructs using socioscientific issues. He currently teaches courses in the graduate programs for both secondary science and curriculum and instruction. He recently served as the Principal Investigator on an NSF-funded Robert Noyce Teacher Scholarship Program grant to support exemplary students with undergraduate degrees in biology in earning their master's degrees in secondary education, along with a teaching endorsement to support students whose first language is other than English.

Goria Campbell-Whatley, PhD, is a Professor in the Department of Special Education and Child Development at the University of North Carolina at Charlotte. Her specialty is STEM, infusing varied cultures into higher education and K-12 curriculum. In addition, she offers solutions for behavior problems, multi-tiered support systems, and social skills training in public schools. Dr. Campbell-Whatley has written numerous articles related to inclusion and has written five books.

Thomas Chandler, PhD, is the Managing Director and Research Scientist at the National Center for Disaster Preparedness and an Associate Member of the Columbia Climate School faculty, Columbia University. He focuses on post-disaster housing and economic recovery, mass care community sheltering and relocation assistance, pandemic preparedness and response, geographic and social networks, and community preparedness. He has 10 years of experience conducting professional development courses pertaining to Geographic Information Systems (GIS) and hazard-resilient housing.

Ting-Han Chang, PhD, is an Assistant Professor in the Department of Educational Leadership and Professional Studies in the College of Education at William Paterson University of New Jersey. Her research interests include advancing equitable STEM educational pathways for first-generation, lower-income students, and/or students of color. She is the Co-Principal Investigator for a multi-year, multi-methods research project, the Student Engagement, Exploration, and Development in STEM (SEEDS) program, a culturally responsive, intergenerational STEM mentorship program with a gaming-based learning component aimed at promoting equity for minoritized middle school and college students.

Hilarie Davis, EdD, is a Curriculum Specialist for NASA Heliophysics Education Activation Team (NASA HEAT). As a former Director of Curriculum in Madison, New Jersey, she is keenly interested in topics like heliophysics that integrate disciplines, provide investigative opportunities for students, and help them develop powerful knowledge. She is a professional evaluator who uses collaborative methods to help teams reflect on their processes and document their outcomes.

Josh DeVincenzo, EdD, is the Assistant Director for Education and Training and Adjunct Assistant Professor at the National Center for Disaster Preparedness, Columbia Climate School, Columbia University. He recently received his doctorate from Columbia University Teachers College with his dissertation on "Enhancing Capacity in Adult Climate Literacy: Investigating Sustainability Mindsets in the U.S. Emergency Management Profession." He focuses on developing training and professional development projects that navigate housing, economic recovery, mass care and sheltering, pandemic planning, and climate literacy.

Shing F. Fung, PhD, is a Research Astrophysicist in the Ionospheric, Thermospheric, Mesospheric Physics Laboratory at the NASA Goddard Space Flight Center, Greenbelt, Maryland. Here he is involved in research on the physics of space plasmas, magnetosphere, and ionosphere, with a particular interest in the natural excitation and propagation of radio and plasma waves. Dr. Fung is also actively involved in the citizen science aspects of the Radio JOVE Project.

Chuck Higgins, PhD, is a Professor of Physics and Astronomy at Middle Tennessee State University in Murfreesboro, Tennessee. He is a planetary radio astronomer studying Jovian and solar radio wave emissions and their propagation through space plasmas. Dr. Higgins is a founding member of the NASA-partner citizen science project Radio JOVE designed to increase science literacy and inspire amateurs to become citizen scientists.

Shuyang Huang, MS, MArch, is a Research Associate working on data management and analysis at the National Center for Disaster Preparedness, Columbia Climate School, Columbia University. She earned a Master of Architecture degree from Columbia's Graduate School of Architecture, Planning, and Preservation (GSAPP) in 2023, and previously obtained an Urban Planning degree there in 2019. Shuyang specializes in developing and researching AI applications incorporated into educational programs.

Jewoong (Jacob) Moon, PhD, is an Assistant Professor of Instructional Technology in the College of Education at the University of Alabama. He earned his PhD degree in Instructional Systems and Learning Technologies at Florida State University, where he was a member of the Cyberlearning Research Group. His expertise is in game-based/immersive (VR, AR, and MR) learning design, instructional design (training, curriculum development), and data analytics integration in technology-enhanced learning.

Hunhui Na, PhD, is an Assistant Professor in the Learning Design and Technology (LDT) program at the University of North Carolina at Charlotte. He is interested in self-directed learning of educators and students and designing technology-enhanced learning experiences for students in STEM+CS. More specifically, his research focuses on how and why early-career teachers participate in online communities for their professional learning, how students develop computational thinking skills in online communities (e.g., Scratch), implementing immersive technologies (VR, AR, gamified content) into classroom contexts for STEM+CS education, and understanding teachers' role in integrating immersive technology within classroom contexts.

Carolyn Y. Ng, MS, is an Education Specialist at NASA Goddard Space Flight Center with the NASA Heliophysics Education Activation Team (NASA HEAT). She and her colleagues have created a framework for heliophysics education designed to guide users at the elementary, intermediate, and advanced levels. Vetted resources are linked to the framework and the Next Generation Science Standards (NGSS). With educators and subject-matter experts, NASA HEAT aims to make the science of understanding our star, the Sun, its interaction with earth and the solar system, and space weather impacts attainable to multiple audiences and learners of all ages.

Yujin Park, PhD, is an Instructional Designer at Georgia Southern University. She earned her PhD in the Instructional Systems and Learning Technologies program at Florida State University. Her research interests include informal teacher professional development in online settings. Her current research interests are STEM teachers' informal professional learning and associations with instructional practices. To explore her research interests, she mainly applies text mining techniques, content analysis, and qualitative case studies.

Tiffany A. Roman, PhD, is an Associate Professor of Instructional Technology at Kennesaw State University in the School of Instructional Technology and Innovation. Her scholarship examines how the design and development of learning technologies, technology-enhanced instructional practices, and active learning environments support student engagement, improve student learning outcomes, and address issues of social justice. Dr. Roman's research interests also include education, broadly conceived, as well as teacher professional development in K-12 STEM education.

Dana Smerda-Mason, EdD, is an Assistant Professor at New Jersey City University in the Educational Technology Department. She has 18 years of experience in K-12 education, where she served as a music educator and technology facilitator. She has worked in global settings as an educator in both China and Guyana, and continues to focus her research on supporting teaching and learning through innovative professional development for educators to help inspire creativity through 21st-century learning engagements.

Shirsh Lata Soni, PhD, is an Early-Career Researcher dedicated to the field of solar physics. Her research interests entail conducting fundamental scientific research in solar physics and space weather using observations and modeling; additionally, her work contributes to the planning and designing of heliophysics missions. Dr. Soni is dedicated to mentorship and outreach, guiding high school and undergraduate students in space science education through NASA-sponsored programs.

Woonhee Sung, EdD, is an Assistant Professor in the School of Education at the University of Texas at Tyler. She earned her EdD in Instructional Technology and Media from Teachers College, Columbia University. Her work encompasses computational thinking, STEM education, and technology integration in educational settings. Her current research focuses on artificial intelligence in education, particularly how AI tools influence K-12 teaching and learning and how AI literacy can be integrated into K-16 education.

Bekir Yıldırım, PhD, earned his undergraduate degree in Science Education from the Faculty of Education at Gazi University, and his master's degree in Biology Education and his PhD in Science Education at the Institute of Educational Sciences at Gazi University. He has authored numerous works on STEM education. Between 2023 and 2024, he completed a one-year postdoctoral fellowship at the University of Minnesota. He is currently a faculty member in the Department of Science Education at the Faculty of Education, Muş Alparslan University. Additionally, he serves as the Deputy Dean of the Faculty of Dentistry at Muş Alparslan University.

1

SETTING THE STAGE

Supporting STEM Teachers Through Online Continuing Professional Development in a Diverse and Digitally Connected World

Heejung An and Ting-Han Chang

Introduction

Science, technology, engineering, and mathematics (STEM) education has gained significant traction in K-12 educational communities in recent years. However, the term "STEM" is often used ambiguously and carries different meanings in different contexts. For clarity and consistency, we begin this chapter by referencing Gonzalez and Kuenzi (2012), who proposed that "STEM education" refers to "teaching and learning in the fields of science, technology, engineering, and mathematics. It typically includes educational activities across all grade levels – from pre-school to post-doctorate – in both formal (e.g., classrooms) and informal (e.g., afterschool programs) settings" (p. 1).

Recently, the field has been promoting the so-called integrative STEM education, which moves beyond teaching each discipline in isolation. Consistent with this trend, in this chapter, we adopt the term "integrative STEM education" to emphasize its interdisciplinary, real-world, and problem-based approaches, while acknowledging that some scholars use the term "integrated STEM" (e.g., Kelley & Knowles, 2016). Proponents emphasize the necessity of unifying the STEM disciplines through integrative approaches and applying knowledge from multiple domains to teach students how to solve complex problems, answer meaningful questions, and explore real-world themes. In fact, the U.S. Department of Education Office of Innovation and Improvement (2016) has promoted integrative STEM education across K-12 teaching and learning, emphasizing the importance of interdisciplinary approaches over traditionally siloed instructions. Such integrative teaching is essential not only to prepare students for the demands of the 21st century but also to ensure national competitiveness in an increasingly interconnected global economy.

Similarly, the release of the Next Generation Science Standards (NGSS) (NGSS Lead States, 2013) has played a significant role in advancing integrative STEM education by fostering interdisciplinary STEM teaching and promoting real-world, inquiry-based learning experiences. The NGSS's three-dimensional learning framework, which integrates (a) science and engineering practices (SEPs); (b) crosscutting concepts (CCCs); and (c) disciplinary core ideas (DCIs), enables students to learn about DCIs while engaging in SEPs – practices that scientists and

DOI: 10.4324/9781003530343-1

engineers use and think about in a broader context. Specifically, the eight SEPs specified in NGSS allow students to actively engage in doing science through student-centered, inductive teaching methods (e.g., project-based, inquiry-based, and guided discovery learning) (Bybee, 2013; Prince & Felder, 2007), which are different from traditional teacher-centered, memorization-based methods. Additionally, the rise of emerging technologies has increased the focus on computing education (Grover & Pea, 2013; Wing, 2017). As a result, NGSS embeds computational thinking within the SEPs, and encourages the use of simulations, modeling, and algorithmic approaches across DCIs. By leveraging tools supported by emerging technologies, educators can create NGSS-aligned learning experiences to support student learning in analyzing complex data, building and testing models, and engaging in authentic problem-solving – all of which are the utilities of integrative STEM teaching that incorporates computing education into the broader vision of 21st-century STEM learning.

In short, at the heart of integrative STEM education is the drive to transcend isolated subject-matter instruction and engage students in interdisciplinary problem-solving rooted in real-world complexities (U.S. Department of Education, 2016). For example, Kelley and Knowles (2016) view STEM education as a holistic approach to teaching the content of two or more STEM domains that fully integrate and connect science, technology, mathematics, and engineering related to real-life problems to enhance learning experiences.

Despite its growing importance, however, successful implementation of integrative STEM education remains limited, with several barriers. A key barrier is that not all schools have adequate facilities and the necessary teaching materials and resources, including qualified teachers. Relatedly, science instruction is often neglected in favor of rigid curricula focused primarily on mathematics and English language arts/literacy to meet standardized testing requirements (Roberts, 2024). Thus, despite an emerging national conversation on promoting integrative STEM teaching, its implementation in many public schools, particularly those in low-income communities, remains inconsistent and underdeveloped (Bybee, 2013; Peters-Burton et al., 2014). For example, many schools serving high-minority or high-poverty populations suffer from a shortage of qualified K-12 STEM teachers. Moreover, many elementary teachers did not major in STEM fields, limiting their confidence or the preparation necessary to adopt an interdisciplinary STEM approach for effective teaching (Brand, 2020; Lin et al., 2021; Lo, 2021; Rich et al., 2018). As a result, elementary teachers often report feeling unprepared to teach science, mathematics, and engineering as separate disciplines, much less integrating these subjects within a cohesive STEM approach in their teaching (Hammack & Ivey, 2017).

Relatedly, despite growing calls for integrative STEM instruction and the aforementioned challenges facing teachers, professional development opportunities remain limited, especially those that effectively support integrative STEM teaching (Bartels et al., 2019). This gap highlights the critical need to develop high-quality continuing professional development (CPD) programs for all educators across K-12 settings through various formats and modalities to develop their competency in applying effective integrative STEM teaching (National Academies of Sciences, Engineering, and Medicine, 2015). A major solution may lie in online CPD, which has become increasingly accessible due to technological advancements with flexible, cost-effective training that transcends geographical limitations (Li et al., 2025). The COVID-19 pandemic further accelerated the adoption and demand for online CPD. In sum, the contemporary contexts suggest a timely need for developing intentional online CPD to foster K-12 educators' capacity to apply integrative STEM teaching in their instruction.

Online Continuing Professional Development in STEM Teaching

Overview of Online CPD

Professional development (PD) typically refers to specific activities or experiences, such as one-time seminars or workshops. However, CPD encompasses the broader, ongoing process of maintaining and enhancing professional knowledge and competence over time. In theory, PD is a component of CPD, however, in practice, many institutions and school districts use the two terms interchangeably or refer to all forms of training as CPD, often without a structured system for tracking or reflection.

In this chapter, we use Guskey's (2000) definition of CPD, which describes "processes and activities designed to enhance the professional knowledge, skills, and attitudes of educators so that they might, in turn, improve the learning of students" (p. 16). CPD is essential for ensuring that teachers continuously maintain and update the knowledge and skills necessary to implement best practices in their profession (Lander et al., 2020). Further, CPD is widely recognized as a critical component of effective instructional practice, since well-prepared and effective teachers have a significant influence on student learning and educational outcomes (Guskey, 2000). As such, CPD offers numerous benefits for teachers, schools, and districts (Borghouts et al., 2021; Guskey, 2000).

Online CPD can take many forms, including degree-based graduate courses, PD programs, and other short types of professional learning opportunities, including various short- and long-term offerings such as workshops, seminars, professional conferences, and other targeted training programs. It may be offered through a range of modalities – synchronous, asynchronous, bichronous (a combination of synchronous and asynchronous formats), or blended/hybrid (a mix of in-person and online), depending on the topic and the intended learning goals (Mohammadi, 2024).

Table 1.1 summarizes different types of online CPD by their respective structures and purposes, whereas Table 1.2 presents a summary of types of online CPD by delivery modality.

Potential Benefits of Online CPD

Among multiple potential benefits of online CPD, the National Research Council (2007) emphasized the following four: (a) flexibility and versatility; (b) opportunities to build community among teachers and across groups; (c) new possibilities for accountability; and (d) improvement of teacher retention by enabling teachers to become more directly involved in their own learning and professional growth. Furthermore, Kabilan (2004) found five aspects of teacher competencies that especially benefited from online CPD: (a) motivation; (b) skills, knowledge, and ideas; (c) self-directed learning; (d) interactive competence; and (e) computer technology awareness and skills. In short, online CPD offers flexibility and opportunities to foster teachers' motivation, competence, and retention.

Evidence of the effectiveness of this type of PD is found by teachers demonstrating stronger skills in collaborating with colleagues both inside and outside their school, showing greater ability to reflect on and evaluate their own practice, gaining more confidence in their teaching, and applying strategies learned through online CPD in ways that positively impact student learning outcomes (Bragg et al., 2021). In short, online CPD provides greater accessibility and flexibility in fostering teachers' competence in applying integrative

TABLE 1.1 Types of Online CPD by Structure and Purpose

Structure and Purpose	*Description/Examples*
Degree-Based Programs	Online master's or doctoral degrees (e.g., M.Ed. in Educational Leadership).
Certificate Programs	Non-degree structured learning paths (e.g., Online STEM or ESL Certificate).
Micro-Credentials/Digital Badges	Competency-based modules with proof of mastery (e.g., Digital Promise).
Massive Open Online Courses (MOOCs)	Free or low-cost online courses available to a large number of participants globally (e.g., Coursera, edX). Access can vary based on country, platform, and course provider policies.
Facilitated Online Courses	Instructor-led, scheduled courses with peer interaction and feedback.
Self-Paced Online Courses	Independent, flexible learning with no real-time components. This can be either a single module or multiple modules, depending on the course design.
Webinars/Virtual Workshops	Real-time or recorded sessions on targeted topics (e.g., Zoom PD sessions). These can be one-time or a series of webinars or virtual workshops.
Online Professional Learning Communities (PLCs)	Collaborative learning through virtual groups.
Virtual Coaching/Mentoring	Personalized support and feedback delivered online.

TABLE 1.2 Types of Online CPD by Delivery Modality

Delivery Modality	*Description/Examples*
Synchronous	Real-time sessions via video conferencing such as Zoom, Microsoft Teams, and Google Meet (e.g., live courses, webinars, and coaching meetings).
Asynchronous	Flexible, on-demand learning (e.g., self-paced courses, recorded lectures, and forums).
Bichronous	Combination of synchronous and asynchronous formats (e.g., live video conferencing sessions + independent modules).
Blended/Hybrid	Mix of online and in-person components (if applicable to the context).

STEM approaches in their teaching. However, specific strategies for maximizing teachers' competence have not been fully presented in current scholarship. To fill this gap, this and other chapters of this book provide research-informed evidence and practical examples in supporting the readers in developing a more comprehensive knowledge of online CPD and integrative STEM education.

Challenges in Designing Effective STEM-Focused Online CPD

While there is a growing demand for and recognition of the potential benefits of online CPD, designing effective STEM-focused online CPD remains a significant challenge. To begin with, a unique set of requirements specific to the STEM disciplines often involves inductive teaching methods (e.g., inquiry-based learning, design-based learning, and project-based learning), visual representations, and hands-on experiences. As a result, transitioning STEM CPD to an online modality requires careful planning, a deep understanding of how technology can be used effectively, and knowledge of appropriate tools for instructional design in an online context. Thoughtful facilitation strategies are also essential for ensuring engagement and learning in virtual environments.

Despite the fact that a number of instructional designs have been developed for online learning, many of these principles were not specifically created for online STEM education in K-12 settings. As a result, they lack the specificity needed to address STEM-related content, pedagogies, and facilitation strategies in online CPD. Furthermore, existing models and frameworks have rarely been applied to the design of online CPD for STEM teacher education.

Suggested Design Principles for Effective Online CPD in Integrative STEM Education

Recognizing the aforementioned gaps, the National Science Foundation's (NSF) DRK-12 program, through its Community for Advancing Discovery Research in Education (CADRE) network (2017), examined 26 projects focused on online or blended PD in K-12 STEM education. Eleven of these projects provide evidence of emerging design principles in three key areas: (a) motivating and sustaining participant engagement; (b) creating opportunities for teachers to collaborate as learners; and (c) supporting reflection on content and practice. Besides outlining both opportunities and challenges of implementing these principles, the CADRE network report also emphasizes the need for continued research to refine and expand effective online CPD practices to guide researchers, instructors, and education leaders in designing and facilitating impactful online CPD programs for integrative STEM teaching. Additionally, Zhou et al. (2023) highlighted three essential components in designing effective STEM-focused online CPD: (a) targeted training in STEM-specific pedagogy; (b) multifaceted and well-structured designs; and (c) careful consideration of PD timing to align with teachers' instructional needs.

Along with these guiding principles and suggestions, the chapters in this book present evidence-based models and strategies for providing effective online CPD for STEM teachers (see Table 1.3). Specifically, the models and strategies were intentionally developed to address one or multiple NSF principles, with practical examples tailored for designing STEM instruction also addressing the unique needs of STEM instruction in online environments.

Diversity, Equity, Inclusion, and Accessibility in Online CPD for STEM Educators

Within the context of planning online CPD for STEM teachers to effectively and meaningfully teach their students, it is important to acknowledge a number of structural inequities related to the needs, challenges, and opportunities in supporting STEM teachers' pedagogy and advocacy for diversity, equity, inclusion, and accessibility (DEIA).

Recognized systemic inequities in STEM education include gender, racial/ethnic, linguistic, cultural, and socioeconomic disparities (U.S. Department of Education, 2016; Whitcomb & Singh, 2021). For example, the underrepresentation of women and non-Whites in STEM degree attainment and careers suggests systemic injustices that limit these groups' access, persistence, and success in the field. According to NSF, in 2021, White individuals comprised 63% of STEM workers in the United States. Additionally, only 18% of women held STEM occupations, compared to 30% of men (NSF, 2024).

TABLE 1.3 Models, Frameworks, and Approaches Discussed in the Book

Model/Framework/Approach	*Definition*	*Representative Reference*
ATTACK (Application, Teaching Strategy, Trial/Try-Out, Assessment, Cognitive Reflection, and Keep/Retain/Generalize) Model	• The ATTACK Model is an instructional design and teaching framework that organizes the learning process into six sequential phases: • Application • Teaching Strategy • Trial/Try-Out • Assessment • Cognitive Reflection and • Keep/Retain/Generalize. • It is designed to guide teachers in structuring lessons that move learners from initial exposure to knowledge through active practice, reflection, and long-term retention.	Campbell-Whatley et al. (2021)
Authentic Learning Model	• The Model of Authentic Learning is an instructional design framework that emphasizes engaging learners in meaningful tasks situated within real-world contexts. • It is structured around nine key elements: • Authentic contexts • Authentic tasks • Access to expert performances • Multiple perspectives • Collaborative knowledge construction • Reflection • Articulation • Coaching and scaffolding • Authentic assessment.	Herrington (2006) and Herrington and Oliver (2000)
Citizen Science (or Participatory Science) Approach	• Citizen Science (or Participatory Science) is a research approach in which members of the public actively engage in scientific inquiry alongside professional researchers. For example, participants may help design studies, collect and analyze data, interpret results, or apply findings, depending on the level of participation.	National Academies of Sciences, Engineering, and Medicine (2018)
Five Ps (Purpose, Place, Platform, Population Interaction Structure, and Profit Model) Framework	• The 5Ps framework is a conceptual model for analyzing and designing virtual communities. • It identifies five interrelated dimensions (Purpose, Place, Platform, Population Interaction Structure, and Profit Model) that together explain how online communities emerge, function, and sustain themselves over time.	Porter (2004)

(Continued)

TABLE 1.3 (Continued)

Model/Framework/Approach	*Definition*	*Representative Reference*
Flipped Learning	• Flipped Learning is a pedagogical approach in which the traditional sequence of instruction is inverted; that is, direct instruction (e.g., lectures) happens outside the classroom, often via videos and readings, while classroom time is reserved for active, student-centered learning such as problem-solving, discussions, labs, or projects.	Flipped Learning Network (FLN) (2014) and Lee (2023)
R2D2 (Read, Reflect, Display, and Do) Model	• The R2D2 Model organizes learning activities into four phases (Read, Reflect, Display, and Do) to address different learning preferences, promote active engagement, and scaffold deeper understanding. • It is especially useful in online education because it provides a structured way to design inclusive and interactive online learning experiences.	Bonk and Zhang (2006)

TMSR (Thinkering, Making, Sharing, and Reflecting) Model	• The TMSR Model is an instructional design model that emphasizes hands-on, inquiry-driven, and collaborative learning through four interconnected phases:
	  • Thinkering
	  • Making
	  • Sharing
	  • Reflecting.
	• It encourages learners to explore ideas, create artifacts, share knowledge and experiences, and critically reflect on their experiences.
		An and Sung (2024)

TPACK (Technological Pedagogical Content Knowledge) Framework	• The TPACK framework is a conceptual model that explains the kinds of knowledge teachers need to effectively integrate technology into their teaching. • It builds on Shulman's (1986, 1987) idea of Pedagogical Content Knowledge (PCK) by adding Technological Knowledge (TK), thereby emphasizing that use of technology in education requires understanding the intersections among content, pedagogy, and technology.	Mishra and Koehler (2006)

In the K-12 STEM education context, White teachers have been overrepresented despite the increasingly diverse student populations around the nation (NSF, 2021). According to the U.S. Department of Education's report from the 2017–2018 National Teacher and Principal Survey (NTPS) (National Center for Education Statistics, 2020), 79% of public school teachers were White. Based on other statistics from NSF (2021), in 2017–2018, 80% of public middle and high school mathematics and science teachers were White, regardless of their school's minority student enrollment. Together, these statistics suggest a need for more intentional efforts to diversify the teaching workforce. Relatedly, scholars have begun calling for a culturally responsive pedagogy to meet the needs of diverse students in STEM education (e.g., Ogodo, 2024).

Given this background, online CPD can play a crucial role in advancing DEIA in professional learning about integrative STEM education, as teachers must be equipped with the knowledge, skills, and leadership capacities necessary to address inequities in their curricula, pedagogy, and mentorship of diverse students pursuing STEM pathways. To provide a more comprehensive understanding in current scholarship, some studies have begun to focus explicitly on DEIA within STEM teacher PD (e.g., Christensen & Knezek, 2024; Litke et al., 2025). Several chapters in this book also place a strong emphasis on DEIA in discussing culturally responsive STEM education, gender-equitable STEM teaching, different modality preferences, and accommodation of multilingual teachers.

Conclusion

The evolving landscape of STEM education underscores the importance of high-quality, accessible, and inclusive CPD for educators. As K-12 schools shift toward integrative and interdisciplinary teaching approaches using technology, teachers need sustained support to build both content knowledge and pedagogical expertise.

Online CPD offers significant promise in meeting these needs by providing flexible, scalable, and personalized learning experiences that can reach educators across diverse contexts. However, designing effective STEM-focused online CPD presents a unique set of challenges. Specifically, the project-based, hands-on, and visual nature of STEM instruction demands carefully developed instructional design models that leverage digital tools while maintaining active engagement and authentic learning experiences. Additionally, many underserved schools still lack access to modern technology and effective strategies for integrating digital tools into instruction (An & Sung, 2023; An et al., 2022). Therefore, it is crucial to raise instructors' and instructional designers' awareness of free digital tools (e.g., open educational resources [OER]) in their design and delivery of online CPD. This book aims to provide practical guidance on how instructors can effectively integrate various strategies into their online CPD. To ensure true impact, online CPD should prioritize DEIA, address systemic inequities that persist in STEM education, and equip teachers with essential knowledge and skills to create inclusive learning environments in their own classrooms.

Moving forward, scholars and practitioners together must continue to refine research-based design principles tailored to STEM-focused online CPD. Intentional partnerships among educators, researchers, and policymakers are essential to create sustainable models that align with teachers' instructional needs while advancing national priorities for STEM education. Ultimately, sustainable and well-designed online CPD can serve as a catalyst for preparing a diverse, highly skilled STEM teaching workforce capable of empowering all students to thrive in the 21st century.

KEY TAKEAWAYS

- High-quality, accessible online CPD is critical for equipping teachers with STEM-specific content and pedagogy, thereby building teacher competency and ensuring consistent implementation across diverse K-12 contexts.
- Designing effective STEM-focused online CPD requires careful attention to the inductive teaching methods, hands-on, and visual nature of STEM learning, supported by well-structured instructional design models tailored to online environments.
- Embedding DEIA in online CPD is essential for addressing systemic inequities in STEM education and preparing teachers to deliver inclusive instruction to an increasingly diverse student population.

References

An, H., Mongillo, G., Sung, W., & Fuentes, D. (2022). Factors affecting online learning during the COVID-19 pandemic: The lived experiences of parents, teachers, and administrators in U.S. high-needs K-12 schools. *The Journal of Online Learning Research (JOLR)*, *8*(2), 203–234. https://www.learntechlib.org/primary/p/220404/

An, H., & Sung, W. (2023). Strategies to help for administrators, teachers, and parents to achieve equitable digital learning in U.S. high-needs school communities. In H. An & D. Fuentes (Eds.), *Digital learning in high-needs schools: A critical approach to technology access and equity in preK-12* (pp. 28–39). Routledge.

An, H., & Sung, W. (2024). The thinkering, making, sharing, and reflecting (TMSR) model: An instructional design model for online collaborative project-based learning. *Educause Review*. https://er.educause.edu/articles/2024/2/thinkering-making-sharing-and-reflecting-a-model-for-online-collaborative-project-based-learning

Bartels, S. L., Rupe, K. M., Lederman, J. S., & Lederman, N. G. (2019). Elementary teachers' concerns about implementing integrated STEM lessons. *School Science and Mathematics*, *119*(6), 339–349.

Bonk, C. J., & Zhang, K. (2006). Introducing the R2D2 model: Online learning for the diverse learners of this world. *Distance Education*, *27*(2), 249–264. https://doi.org/10.1080/01587910600789670

Borghouts, L., Slingerland, M., Weeldenburg, G., van Dijk-van Eijk, B., Laurijssens, S., Remmers, T., & Haerens, L. (2021). Effectiveness of a lesson study intervention on teacher behaviour and student motivation in physical education lessons. *Physical Education and Sport Pedagogy*, *28*(2), 121–138.

Bragg, L. A., Walsh, C., & Heyeres, M. (2021). Successful design and delivery of online professional development for teachers: A systematic review of the literature. *Computers & Education*, *166*, 1–23.

Brand, B. R. (2020). Integrating science and engineering practices: Outcomes from a collaborative professional development. *International Journal of STEM Education*, *7*, 1–13. https://doi.org/10.1186/s40594-020-00210-x

Bybee, R. W. (2013). *The case for STEM education: Challenges and opportunities*. NSTA Press.

Campbell-Whatley, G. D., Rodriguez, D., & Agrawal, J. (2021). *STEAM meets story: Using adolescent fiction and film to spark deeper learning*. Teachers College Press.

Christensen, R., & Knezek, G. (2024). Impacting student perceptions of engagement and student voice through equity-focused teacher professional development in a simulated teaching environment. *Journal of Interactive Learning Research*, *35*(2), 141–166.

Community for Advancing Discovery Research in Education. (2017). *Emerging design principles for online and blended teacher professional development in K-12 STEM education*. Education Development Center, Inc. Retrieved from https://cadrek12.org/resources/emerging-design-principles-online-and-blended-teacher-professional-development-k-12-stem

Flipped Learning Network (FLN). (2014). *The four pillars of F-L-I-P™* (Reproducible PDF may be found at www.flippedlearning.org/definition).

Gonzalez, H. B., & Kuenzi, J. J. (2012). *Science, technology, engineering, and mathematics (STEM) education: A primer*. Congressional Research Service. https://sgp.fas.org/crs/misc/R42642.pdf

Grover, S., & Pea, R. (2013). Computational thinking in K-12: A review of the state of the field. *Educational Researcher*, *42*(1), 38–43.

Guskey, T. R. (2000). *Evaluating professional development*. Corwin Press.

Hammack, R., & Ivey, T. (2017). Elementary teachers' perceptions of engineering and engineering design. *Journal of Research in STEM Education*, *3*(1/2), 48–68. https://doi.org/10.51355/jstem.2017.29

Herrington, J. (2006, October 13–17). *Authentic e-learning in higher education: Design principles for authentic learning environments and tasks* [Paper presentation]. World Conference on E-Learning in Corporate, Government, Healthcare, and Higher Education (E-Learn 2006).

Herrington, J., & Oliver, R. (2000). An instructional design framework for authentic learning environments. *Educational Technology Research and Development*, *48*(3), 23–48. https://doi.org/10.1007/BF02319856

Kabilan, M. K. (2004). Online professional development: A literature analysis of teacher competency. *Journal of Computing in Teacher Education*, *21*(2), 51–57.

Kelley, T. R., & Knowles, J. G. (2016). A conceptual framework for integrated STEM education. *International Journal of STEM Education*, *3*(1), 1–11. https://doi.org/10.1186/s40594-016-0046-z

Lander, N., Lewis, S., Nahavandi, D., Amsbury, K., & Barnett, L. M. (2020). Teacher perspectives of online continuing professional development in physical education. *Sport, Education and Society*, *27*(4), 434–448. https://doi.org/10.1080/13573322.2020.1862785

Lee, J. (2023). Flipped learning. In O. Zawacki-Richter & I. Jung (Eds.), *Handbook of open, distance and digital education* (pp. 1179–1196). Springer. https://doi.org/10.1007/978-981-19-2080-6_69

Li, Z., Wang, C., & Bonk, C. J. (2025). Generative AI for teachers' self-directed professional development: A mixed-methods study. *TechTrends*. https://doi.org/10.1007/s11528-025-01123-8

Lin, P. Y., Chai, C. S., & Jong, M.S.Y. (2021). A study of disposition, engagement, efficacy, and vitality of teachers in designing science, technology, engineering, and mathematics education. *Frontier in Psychology*, *12*, 1–12. https://doi.org/10.3389/fpsyg.2021.661631

Litke, E., Wilson, J., & Hill, H. C. (2025). Equity-focused, rubric-based coaching: An incremental improvement approach to supporting teachers to shift toward more equitable mathematics instruction. *Education Sciences*, *15*(4), 1–18.

Lo, C. K. (2021). Design principles for effective teacher professional development in integrated STEM education: A systematic review. *Educational Technology & Society*, *24*(4), 136–152. https://www.jstor.org/stable/48629251

Mishra, P., & Koehler, M. J. (2006). Technological pedagogical content knowledge: A new framework for teacher knowledge. *Teachers College Record*, *108*(6), 1017–1054.

Mohammadi, G. (2024). Teachers' CALL professional development in synchronous, asynchronous, and bichronous online learning through project-oriented tasks: Developing CALL pedagogical knowledge. *Journal of Computers in Education*, *11*, 401–422. https://doi.org/10.1007/s40692-023-00260-4

National Academies of Sciences, Engineering, and Medicine. (2015). *Science teachers' learning: Enhancing opportunities, creating supportive contexts*. The National Academies Press. https://doi.org/10.17226/21836

National Academies of Sciences, Engineering, and Medicine. (2018). *Learning through citizen science: Enhancing opportunities by design*. The National Academies Press. https://doi.org/10.17226/25183

National Center for Education Statistics. (2020). *Data point: Race and ethnicity of public school teachers and their students*. https://nces.ed.gov/pubs2020/2020103/index.asp

National Research Council. (2007). *Enhancing professional development for teachers: potential uses of information technology: Report of a workshop*. The National Academies Press. https://doi.org/10.17226/11995

National Science Foundation (NSF). (2021). *Elementary and secondary STEM education*. https://ncses.nsf.gov/pubs/nsb20211/table/K12-4

National Science Foundation (NSF). (2024). *The STEM labor force: Scientists, engineers, and skilled technical workers*. https://ncses.nsf.gov/pubs/nsb20245/representation-of-demographic-groups-in-stem#:~:text=In%202021%2C%20Black%20or%20African,with%2018%25%20of%20all%20workers

NGSS Lead States. (2013). *Next Generation Science Standards: For states, by states*. The National Academies Press.

Ogodo, J. A. (2024). Culturally responsive pedagogical knowledge: An integrative teacher knowledge base for diversified STEM classrooms. *Education Sciences*, *14*(2), 1–13. https://doi.org/10.3390/educsci14020124

Peters-Burton, E. E., Lynch, S. J., Behrend, T. S., & Means, B. B. (2014). Inclusive STEM high school design: 10 critical components. *Theory Into Practice*, *53*(1), 64–71. http://dx.doi.org/10.1080/00405841.2014.862125

Porter, C. E. (2004, November 1). A typology of virtual communities: A multi-disciplinary foundation for future research. *Journal of Computer-Mediated Communication*, *10*(1). https://doi.org/10.1111/j.1083-6101.2004.tb00228.x

Prince, M., & Felder, R. (2007). The many faces of inductive teaching and learning. *Journal of College Science Teaching*, *26*(5), 14–20.

Rich, P., Belikov, O., Yoshikawa, E., & Perkins, M. (2018). Enablers and inhibitors to integrating computing and engineering lessons in elementary education. *Journal of Technology and Teacher Education*, *26*(3), 437–469.

Roberts, M. J. (2024). *Closing the gap: Addressing the deficiency of STEM professional development for elementary educators* [Unpublished doctoral dissertation]. University of Pittsburgh.

Shulman, L. (1986). Those who understand: Knowledge growth in teaching. *Educational Researcher*, *15*(2), 4–14.

Shulman, L. S. (1987). Knowledge and teaching: Foundations of the new reform. *Harvard Educational Review*, *57*(1), 1–22.

U.S. Department of Education, Office of Innovation and Improvement. (2016). *STEM 2026: A vision for innovation in STEM education*. https://www.ed.gov/sites/ed/files/2016/09/AIR-STEM2026_Report_2016.pdf

Whitcomb, K. M., & Singh, C. (2021). Underrepresented minority students receive lower grades and have higher rates of attrition across STEM disciplines: A sign of inequity? *International Journal of Science Education*, *43*(7), 1054–1089. https://doi.org/10.1080/09500693.2021.1900623

Wing, J. M. (2017). Computational thinking's influence on research and education for all. *Italian Journal of Educational Technology*, *25*(2), 7–14.

Zhou, X., Shu, L., Xu, Z., & Padrón, Y. (2023). The effect of professional development on in-service STEM teachers' self-efficacy: A meta-analysis of experimental studies. *International Journal of STEM Education*, *10*(37), 1–20. https://doi.org/10.1186/s40594-023-00422-x

SECTION I

Designing Effective Online Continuing Professional Development for STEM Educators: Models, Strategies, and Case Examples

2

PROFESSIONAL DEVELOPMENT PATHWAYS THAT SUPPORT INCLUSION AND INNOVATION

A STEAM Education Approach to Virtual Educator Training

Dana Smerda-Mason

Introduction

Despite the promising results of the science, technology, engineering, and mathematics (STEM)/science, technology, engineering, arts, and mathematics (STEAM) instructional model, its widespread adoption is hindered by challenges related to implementation, including insufficient training, significant variability in both understanding and application of STEAM principles, limited time for interdisciplinary collaboration, and lack of institutional support (Land, 2013). Indeed, these are the very factors that have been identified as essential enablers of successful implementation.

Continuing professional development (CPD) that incorporates a virtual learning community can provide a practical solution to these challenges, including targeted professional development and development of a robust support network among educators where they can receive the support they need to upskill professionally. To that end, this chapter focuses on Porter's (2004) key attributes that exist in successful virtual communities: purpose, place, platform, population, and profit.

By identifying both the limitations and opportunities of current implementation efforts, this chapter aims to contribute a deeper understanding of how to better support K-12 educators in delivering an integrated, future-ready approach to convergence education in their professional practice.

> The morning bell rings, and the rush of the school day begins with teachers welcoming and preparing students for the day as they bring them into the classroom. In some cases, breakfast is served, and books are packed up for morning classes. A few minutes later, this first group of students leave, and a new batch of students enter, and lessons begin. Educators share well-prepped content that was designed in advance outside of school to ensure they meet the needs of local standards, high-stakes testing, cultural appropriation, content knowledge, differentiation, social and emotional wellness, and 21st-century standards. After managing student needs, addressing discipline issues, ensuring that enduring understandings have been absorbed, teachers soon

DOI: 10.4324/9781003530343-3

> clean up and prepare for their next class to arrive. In a short time that class period is also over, and the students are off to their next period, followed by a fresh group of learners entering.
>
> This pattern repeats all day long, and any time teachers have without students—preparation periods as well as lunch breaks—is often taken up by corresponding with parents and administrators. At the end of the school day, grading and other administrative tasks continue into the evening, in addition to any lesson preparations to ensure that everything is ready for the next day's class schedule.
>
> Such a packed schedule leaves limited time for teachers to continue learning themselves, even though this is an essential part of developing as an education professional. There are usually a few professional development days scheduled throughout the school year on general topics provided by the school district. In addition, if requested, educators may be able to obtain approval to attend a workshop during the day, but given the growing responsibilities facing educators in the 21st century, there is a need for creative approaches to targeted professional learning that meets the specialized need of each professional.

CPD for educators is an essential part of their professional practice, and directly impacts what occurs in the classroom. Reflective of the evolution of society, what qualifies as effective CPD has changed over the past 20 years, as educators are tasked with the unique responsibility of preparing students for future roles within the 21st century. Technology and innovation drive much of this change, and what was previously successful needs to be reevaluated so that the education community can design learning experiences that reflect current and future student and societal needs (Deev & Finogeev, 2023). One successful path toward this goal for educators is to engage in an online virtual community designed to address the broad range of issues they face in the 21st century.

Historically, subject matter has been divided by content and taught individually, but educators are increasingly finding that when knowledge, techniques, and expertise from a variety of domains are integrated and brought together for problem-solving in the classroom, students have a better chance at success (Herr et al., 2019). This is also reflective of the nature of the 21st-century workforce, where experts from different fields collaborate and innovate in a transdisciplinary environment (WIREs Authors, 2019). As discussed in the following section, STEM/STEAM as a content area has been most successful when approached as a *convergence education experience*; that is, when learners study integrated content from various academic subjects.

Virtual Communities Blending Formal and Informal Learning

Historically, professional development has been delivered in person with a focus on content, technological integration, or general pedagogy, but it has often failed to focus on a combined approach to the needs of STEM/STEAM educators (Chai, 2018). Traditional formats that rely on workshops, lectures, and conferences, so-called passive learning, do not always allow for hands-on active learning experiences (Conradty & Bogner, 2020). That is, while more traditional formats can be an effective way of learning for some, educators know from their own students that layering a variety of learning experiences and differentiating instruction can inspire engagement and personal interest in the material at hand.

Virtual communities have the opportunity to take the best of formal and informal learning experiences and blend them with the support of a community of individuals who share similar interests to maximize their engagement in CPD through autonomy and the use of technology. Such online learning spaces can take on a variety of shapes to meet the needs of the community members and their shared area of interest. Ranging from a group of embroidery crafters that want to share and inspire ideas on a Facebook group, to resumé writers and recruiters sharing advice on LinkedIn groups, or educators seeking collaborative connection related to assistive technology (AI) policy through a Google Group forum, there are communities that can meet the needs of almost anyone looking to pursue a given interest.

Members of these egalitarian communities can enter as beginners, and through access to and engagement with experts and resources receive a sense of belonging and encouragement (Wenger, 2009). As these novice members grow and develop, they, in turn, become equipped to share and support others in the community, and in the process benefit from refining their knowledge base by helping others. A powerful aspect of this type of environment is the convenience of accessibility, which has the potential to attract and support members who might otherwise have been precluded from these types of experiences if they were only available at a specific location and required in-person attendance. That is, the online nature of these communities helps active adult learners, especially, meet their needs as they allow members to engage when they need information or when they have time to seek information.

Our understanding of how adults learn can be characterized by the concept of andragogy, which proposes that adult learning is driven by (a) the need to know something, (b) the readiness to learn it, (c) the ability to learn autonomously, (d) intrinsic motivation toward a particular goal, and (e) the opportunity to base these new understanding on past knowledge and experiences (Knapke et al., 2024).

Virtual communities in their many forms naturally satisfy many of these adult learning needs, and can help connect educators with impactful learning that can satisfy their CPD training requirements. Each virtual community can and should be customized to meet the needs of the members it serves. To do this responsibly, it is essential to understand the challenges for which educators require support. In the context of this chapter, this involves helping them develop their skills related to STEM/STEAM education.

Challenges

The introduction to this chapter described a typical day in the life of many educators, highlighting the lack of time available to invest in professional growth. In addition to finding time for CPD, today's educators face a variety of other challenges; for example, it is not uncommon for them to be expected to implement new programs with limited training.

In particular, content knowledge and pedagogical design are areas that require unique support when implementing a convergence education approach to teaching. That is, educators are usually trained in a specific domain, so it can be difficult for them to implement content knowledge across multiple disciplines, as in the case of STEM/STEAM. Even with the opportunity and time to conduct professional research, it can be challenging to discern the efficacy of sources when searching for materials in subjects in which one is not an expert, for example. In a study of Hong Kong educators teaching STEM, only 5.53% reported having a comfort level in carrying out lessons for their students that required a transdisciplinary focus (Geng et al., 2018). It can also be difficult to evaluate new programs that are being integrated into the local curriculum without content knowledge beyond one's own area of expertise, especially when considering the pedagogical approaches that are unique to delivering specialized content in each domain.

Another point to remember in relation to teacher preparation and preparedness is that most educators today did not encounter STEM/STEAM education as a transdisciplinary approach in their own learning experiences as children or as preservice education students. Universities are now starting to address this critical issue by incorporating STEM/STEAM as a methodological offering to preservice education students.

Typically, educators are required to complete a liberal arts program toward a Bachelor of Arts degree to obtain experience and certification to become an educator unless they are planning to teach subject-specific or advanced courses in secondary education (Education-Colleges.com, 2021). In that case, they are required to hold either a Bachelor of Science degree prior to pursuing their education degree, or obtain additional certifications in a particular area to be deemed qualified. Many STEM educators are required to obtain technical training as they pursue a Bachelor of Science track toward their specified domain, leaving them at a disadvantage in terms of being prepared to teach these domains. That is, STEM/STEAM as a teacher preparation program is rare among education major options in higher education, and only recently are STEM/STEAM fundamentals being offered as a distinct course in teacher preparation programs—and then usually only at the graduate level. For these reasons, CPD is an essential part of implementing a 21st-century convergence education approach, and has the potential to make a strong impact on experienced educators.

As part of effective CPD, collegial support through collaborative professional learning communities (PLCs) needs to be established so that educators have set times for planning and discussion as they develop solutions to student learning through the collaboration of a variety of domain experts. Unfortunately, with scheduling challenges and full class loads, it can be difficult for educators to find the time during the regular school day to help one another while attending to their own list of responsibilities. Overcoming these hurdles requires administrative support (Dong et al., 2018). For example, providing class coverage so educators can attend conferences and workshops has been found to be effective, as has bringing in high-quality professional development solutions for in-house training. To date, however, not all administrators are aware of or embrace the benefits of STEM/STEAM integration, so it is not uncommon for educators to be left to navigate these areas of need on their own while juggling the daily pressures related to student achievement through standardized test scores and numerous other responsibilities.

Implementing a transdisciplinary learning model in a school environment often requires adjustment in the culture of the school. Administrators at all levels need to set the tone so that the institution can successfully incorporate support for professional learning, collaboration, adjustments in traditional practices, and develop an understanding of what their educators need to be able to teach students through cross-curricular objectives. This goes beyond training in content knowledge and asking teachers to change their methods. Administrators must have the freedom to lead by incorporating a more holistic approach to learning, such as adjusting school schedules to accommodate how educators collaborate and prepare for teaching with a convergence education approach (McKee, 2022). In short, the collective success of a school implementing a convergence approach to STEM/STEAM requires support from colleagues and administrators alike to help equip educators with resources, class time, appropriate instruction space and storage of materials, and collaborative time with other colleagues (Dong et al., 2018).

Professional Development Solutions for STEM/STEAM Educators

The learning environment of a STEM/STEAM classroom differs from the expected order and function of a traditional 20th-century classroom. For example, students engage in collaborative activities, and often have the autonomy to make their own decisions about how to satisfy the requirements of assignments. They engage in discussion and experimentation, and learn in an active and hands-on environment that sometimes looks like a room full of organized chaos with students buzzing about as they receive coaching and guidance from their teacher, who serves as a facilitator for learning. In these classrooms, students are engaging in a socially constructivist learning experience as they build their understanding of a particular topic (Saleem et al., 2021) while bringing in elements of other subject domains required to bring it to fruition.

To meet the demands of 21st-century classrooms like this, traditionally trained educators need innovative approaches to how CPD is designed and presented. Adult learning settings where educators attend in a seated environment listening to a speaker discuss what has to be done is only partially effective. Instead, educators need to be exposed to collaboration and CPD experiences that allow them to construct their knowledge first-hand, just as their students would (Schaffauser, 2019). Layering a variety of learning experiences such as lecture, small-group collaboration, video, and self-paced activities helps develop a stronger and more enduring understanding of the information at hand (Schaffauser, 2019) and is an effective means of introducing new methods to educators.

Based on a 2022 study, regular educators of STEM/STEAM in New Jersey expressed an interest in learning new material in either a hands-on environment or a completely virtual setting (Smerda-Mason, 2022). As illustrated in the opening to this chapter, educators have no, or limited, time during the school day to engage in professional learning. That is, they use their limited preparation time as efficiently as possible to communicate with parents, grade classwork, and prepare for upcoming classes, barring any emergencies or interruptions that may occur and claim that already small block of free time. This leaves a few in-service days on the school calendar and personal time as options for professional training and upskilling.

When properly designed, the format of virtual learning for professional development can support teachers through the convenience of autonomous learning, as described in the following.

Elements of Virtual Communities as a CPD Support Solution

A virtual community can be designed to meet the unique challenges and specific needs of 21st-century educators interested in developing their skills as STEM/STEAM educators. To that end, Porter's (2004) typology of virtual communities can provide a helpful framework. Specifically, Porter refers to the "five Ps" of key attributes that exist in virtual communities as follows: *purpose*, *place*, *platform*, *population*, and *profit model*. Since each virtual community possesses varied and multifaceted features, this model can help illuminate important areas of consideration for an effective and efficient design solution.

Purpose

The first attribute is the *purpose*—in the current context, a CPD solution for STEM/STEAM educators.

Place

The next attribute is *place*, which in this case was determined to be a completely virtual setting so that PD opportunities are accessible by all teachers at their convenience without the constraints of location or schedule. In other cases, virtual communities include in-person elements or operate in a hybrid format, which for some purposes may be a more effective choice (Porter, 2004). A hybrid format often incorporates elements of in-person activities or interactions, while also providing a virtual space where participants can continue to share information and offer support beyond the structure of the in-person engagements. This sort of flexibility meets the needs of the learners that extend past the limited time that they are able to be in person, in providing answers to questions, ideas for materials.

For example, if this type of virtual community is being utilized in a particular school district, the virtual component of this community could be designed to meet their particular needs that arise after the hours of the school day. When integrating STEM/STEAM education, a space of discussion to share about what is working or not working during in classes after hours can be widely beneficial. Educators developing lesson plans for the following day can explore shared resources from fellow colleagues, and when integrating material from outside of their domain of expertise, they have experienced colleagues with differentiated skills at their disposal to connect with.

The district can also utilize this space to share videos, or create professional learning content to complement other training efforts that occur in person. This does not have to be a place for after-hours only, but can also be accessed during the school day in individual preparation periods or when teacher schedules allow. The flexibility that the hybrid option provides is beneficial not only to educators committed to developing STEM/STEAM education skills, but also to the overall district when having to solve logistical challenges often associated with CPD.

Platform

Platform, the third attribute, speaks to the design of interactions among the members of a specific community. It is important to decide where this space should be hosted, and to find an option that can also be easily accessible for future community members. There are a variety of options to consider, such as social media platforms, wikis, community forums, learning platforms, and websites. For example, if a social media platform is determined to be the best option, choose one that will have greater accessibility for those you are inviting and hope to include over time. A simple Google site will serve as the working example here.

There are many benefits to this choice, the first of which is that a Google site is easy to navigate and provides intuitive and simple templates that help people of all technology levels contribute to the design and upkeep of the community space. Sharing capabilities can be easily regulated to meet the specific needs of each member based on their role. Google is commonly used in learning spaces with students, so many teachers may already be familiar with it. Also, with a simple web search, new users can access countless free tutorial videos to learn the nuances that they seek as they utilize this platform and design a CPD space that meets the customized possibilities of their imagination. The basic model is free, so there is no need for members to make a financial commitment to have access, unless they wish to invest in access to upgraded features. Some of the paid features can include additional storage, advanced management controls and security, advanced AI functions, and access to technical support at any hour.

The website can include areas for posting recommended resources and classroom materials by subject domain, as well as virtual resources. Videos to use with students or tutorials on new techniques or content of interest to teachers can be organized and shared here as well. Templates for assessments and lesson plans can also be shared within the community, making the integration of multiple subjects easier for educators to learn to plan.

These are all meaningful areas that educators can explore in their own, often limited, time. Instead of starting research on a particular topic from scratch, for example, they can quickly browse recommendations from peers in the field, which not only saves time but also allows self-directed and convenient engagement, whether it be 5-minutes between classes or 30-minute when unwinding at the end of the day. As long as they have a connection to the Internet, educators can be thinking and formulating new ideas for their students to experience in the classroom.

Population

When successful practices are achieved, educators can easily share what is working to help support others; conversely, colleagues can share what is not working and receive timely feedback from other subject matter experts. This speaks to the next of Porter's five Ps: population and the pattern of interaction among community members. It is important to include a means of both synchronous and asynchronous communication for community interaction and growth. A forum can be a space developed for educators to engage, and a Google Group can easily be added to offer this for asynchronous discussion. In addition, social media groups on Facebook, LinkedIn, or Twitter/X, to name a few, can be established and linked to the webpage to keep the conversations open and supplement the materials on the website.

Typically, in PD settings, there are clear experts and novices, but due to the convergence of all domains in this model, it becomes a unique community of practice (Wenger-Trayner & Wenger-Trayner, 2015), where educators occupy both roles. For example, each educator serves as an expert in their particular domain but can also quickly receive support and answers related to other domains in which they have less experience. The nature of communication in these types of groups also helps support professional learning beyond the instance at which the educator gathers the information. They have continued support and guidance as they experiment with new methods in their classroom throughout the year.

In addition to resources, materials, videos, and the forum, a special page for CPD should also be included. Here, the community can share links and recommendations of training opportunities and conferences, as well as upcoming webinars from different providers.

For example, unique professional growth opportunities for educators are offered at NASA centers, government agencies such as the Department of Defense, and other organizations that support STEM/STEAM education globally. Such trainings are offered in many cities across the country, and can offer that blend of hands-on active learning that educators crave in their CPD experiences.

The community can also establish sharing opportunities, where experts in the community share their personal knowledge through online webinars for community members. This would meet the need for synchronous learning opportunities and help strengthen the community through communication and collaboration. Using this format, expert members benefit from sharing their expertise as well as being challenged about what they know while helping novice members overcome hurdles of implementation. In short, the virtual community can provide support and growth for educators of all levels.

Profit Model

The final of Porter's five Ps, profit model, refers to how and if the community generates revenue (Porter, 2004). This does not apply to each group, but should be considered for virtual communities that establish a monetary exchange for knowledge, services, or goods. In some cases, even virtual communities that do not have an economic component may consider this element if they sell advertising space on their platform or recognize the potential for something like this as a means of long-term sustainability if the community grows to something that requires maintenance fees or stipends for work beyond what members are volunteer basis.

Table 2.1 illustrates the five elements in Porter's (2004) typology of virtual communities.

TABLE 2.1 Typology of Educators' Virtual Community for STEM/STEAM Continuing Professional Development

Attribute	*Non-profit community*
Purpose	CPD solution for educations in a K-12 STEM/STEAM education environment.
Place	• In-person school environment • Online social platform • Hybrid option, blending the in-person experience with resources and training provided virtually with autonomous access
Platform	Hybrid online platform such as a Google site provides asynchronous access for educators to interact autonomously at their convenience and connect with colleagues, as well as share and rely on posted training and classroom resources.
Population Interaction Structure	• Asynchronous • Autonomous • Option for district-hosted training content available to community members
Profit Model	Non-revenue-generating, but can indirectly save money for the participating school district through free exchange of information and resources, as well as additional class coverage and other costly training requirements involved in providing traditional CPD to teaching staff.

Conclusion

The goal of education is to prepare students to become successful members of contributors to society. But "how do we educate our children to take their place in the economies of the 21st century, given that we can't anticipate what the economy will look like next week?" (Robinson, 2010). The field of education is challenged by this very question and is working at finding solutions to equip our students with the skills they will need to be successful in the 21st century, while at the same time overcoming challenges in pedagogical design in need of change. As society evolves, our teaching practices must also change to match the level of innovation of the changes we are experiencing.

Overcoming the aforementioned challenges is part of a widespread adoption of convergence education and can help instill 21st-century skills as well as divergent problem-solving abilities in future generations (Conradty & Bogner, 2020). Innovative approaches to CPD are essential to the success of this effort. By experiencing this type of learning themselves through a virtual community, educators are afforded the opportunity to step into the shoes of their students and find creative ways to instill impactful learning in a virtual format to take back to their classrooms. This type of virtual learning community includes andragogy practices and supports the professional growth of busy educators who need time to learn autonomously at a time and place that is convenient for them. With such support, they have the opportunity to engage in efficient and meaningful CPD, mentoring and supporting each other, while igniting their passion in new ways to do what they do best: pass civilization on to the next generation.

KEY TAKEAWAYS

- Convergence education is a method that is relied upon for STEM/STEAM, where educators teach more than one subject domain at a time to help layer the learning experience and enhance student engagement and enduring understandings.
- Traditionally, educators study a liberal arts track of learning (BA) to become highly qualified education professionals. STEM/STEAM education relies on many technical fields, the experts of which pursue a BS to achieve their first level of expertise. Educators do not receive training to integrate STEM/STEAM in preservice training and, therefore, require additional continued professional development to adopt these learning models with confidence.
- Virtual communities can be a CPD solution for educators to help them integrate convergence education and receive support from a community of educators with different skillsets.
- A virtual community can be designed as a complement to the in-person experience educators have while trying to integrate new content and methods of STEM/STEAM education in the classroom. Challenges and successes can be shared through this community, discussed and supported, and training resources as well as classroom tools and other forms of learning can be accessed between classes or after school day is over.
- Virtual communities support school districts because they offer a customized and differentiated approach to learning by relying on internal talent to cross-pollinate important skills so that educators can rely on one another to develop strong self-efficacy in STEM/STEAM education.
- Porter's (2004) typology of virtual communities can provide a helpful framework for developing virtual communities by ensuring the key attributes that exist in virtual communities established as the "five Ps," *purpose*, *place*, *platform*, *population*, and *profit model*, are properly considered for an effective and efficient design solution.

References

Chai, C. S. (2018). Teacher professional development for science technology engineering and math (STEM) education: A review from the perspectives of technical pedagogical content. *Asia-Pacific Education Researcher*, *28*(1), 5–13. https://doi.org/10.1007/s40299-018-0400-7

Conradty, C., & Bogner, F. X. (2020). STEAM teaching professional development works: Effects on students' creativity and motivation. *Smart Learning Environments*, *7*(26). https://slejournal.springeropen.com/track/pdf/10.1186/s40561-020-00132-9.pdf

Deev, M., & Finogeev, A. (2023, June). Application of the convergent education model in the development of a smart learning environment. *Telematics and Informatics Reports*, *10*(10005). https://www.sciencedirect.com/science/article/pii/S2772503023000117

Dong, Y., Xu, C., Song, X., Fu, Q., Chai, C. S., & Huang, Y. (2018). Exploring the effects of contextual factors on in-service teachers' engagement in STEM teaching. *Asia-Pacific Education Researcher*, *28*, 25–34. https://doi.org/10.1007/s40299-018-0407-0

Education-colleges.com. (2021). https://www.education-colleges.com/bachelors-degree-ba-bs/#:~:text=All%20states%20in%20the%20U.S.,and%2C%20eventually%2C%20for%20teaching

Geng, J., Jong, M. S., & Chai, C. S. (2018). Hong Kong teachers' self-efficacy and concerns about STEM education. *Asia-Pacific Education Researcher*, *28*, 35–45. https://doi.org/10.1007/s40299-018-0414-1

Herr, D.J.C., Akbar, B., Brummet, J., Flores, S., Gordon, A., Gray, B., & Murday, J. (2019, August). Convergence education—An international perspective. *Journal of Nanoparticle Research*, *21*(229). https://doi.org/10.1007/s11051-019-4638-7

Knapke, J. M., Hildreth, L., Molano, J. R., Shuckman, S. M., Blackard, J. T., Johnstone, M., Koporas, E. J., Lamkin, M. K., Lee, R. C., Kues, J. R., & Mendell, A. (2024, March 28). Andragogy in practice: Applying a theoretical framework to team science training in biomedical research. *British Journal of Biomedical Science*, *81*(12651). https://doi.org/10.3389/bjbs.2024.12651

Land, M. H. (2013). Full STEAM ahead: The benefits of integrating the arts into STEM. *Procedia Computer Science*, *20*, 547–552. https://doi.org/10.1016/j.procs.2013.09.317

McKee, A. (2022). We need to stop teaching high school subjects separately. *K-12 Dive*. https://www.k12dive.com/news/we-need-to-stop-teaching-high-school-subjects-separately/624932/

Porter, C. E. (2004, November 1). A typology of virtual communities: A multi-disciplinary foundation for future research. *Journal of Computer-Mediated Communication*, *10*(1). https://doi.org/10.1111/j.1083-6101.2004.tb00228.x

Robinson, K. (2010, November 4–5). *Keynote address*. NJEA 2010 Convention. Atlantic City, NJ.

Saleem, A., Kausar, H., & Deeba, F. (2021, December). Social constructivism: A new paradigm in teaching and learning environment. *Perennial Journal of History*, *2*(2), 403–421. https://doi.org/10.52700/pjh.v2i2.86

Schaffauser, D. (2019). Getting trained to teach robotics. *THE Journal (Technological Horizons in Education)*, *20*(6). https://thejournal.com/articles/2019/03/28/getting-trained-to-teach-robotics.aspx

Smerda-Mason, D. (2022). *Professional development concerns and solutions in STEAM education* (ProQuest). [Doctoral dissertation, New Jersey City University].

Wenger, E. (2009, September 27). *Communities of practice and social learning systems: The career of a concept*. https://www.wenger-trayner.com/wp-content/uploads/2022/09/09-10-27-CoPs-and-systems-v2.0.pdf

Wenger-Trayner, E., & Wenger-Trayner, B. (2015). *Introduction to communities of practice*. https://wenger-trayner.com/introduction-to-communities-of-practice/

WIREs Authors. (2019, July 29). Are we really stronger together? *Advanced Science News*. https://www.advancedsciencenews.com/a-framework-for-measuring-if-were-really-stronger-together/

3

THE TMSR AND 4-ONS MODEL

Design-Based Learning in Online Continuing Professional Development Integrating Arts in STEM

Woonhee Sung and Heejung An

Introduction

The landscape of professional development has shifted dramatically with the rise of online learning environments, especially post-COVID-19. As a result of the increased need for online courses and programs, online teacher professional development and online continuing professional development (CPD) have grown substantially (Bragg et al., 2021; Lay et al., 2020). However, while online delivery offers broader access and flexibility for all learners, it also presents unique challenges (Chiu et al., 2021) primarily related to limited communication and interactions between learners and instructors, interactions among learners, and feelings of disconnectedness leading to low engagement (Angelino et al., 2007; Bolliger & Inan, 2012).

These challenges, centered on collaboration and engagement in online learning, have raised specific concerns among science, technology, engineering, arts, and mathematics (STEAM)-related disciplines due to the interactive and hands-on nature of these subjects (Quigley, 2016; Taysever, 2016). To overcome issues grounded in online delivery for science, technology, engineering, and mathematics (STEM) subject areas, learning activities need to be based on authentic practices, highlighting learners' involvement throughout the projects, which, in turn, encourages interactions among learners. These unique design principles have been analyzed in several studies (Lo, 2021); however, their feasibility in online delivery has not been studied widely. Moreover, to date, best practices integrating the arts in STEM education that are applicable in online delivery are limited (Wong et al., 2022).

This chapter introduces an engaging online professional development process that adds the "A (Arts)" in STEM education. As noted by Kim and Park (2012), arts-oriented learning activities can promote creative, interdisciplinary, and real-world problem-solving when included in STEM education. The lesson activities introduced in this chapter have intentionally combined music with programming, robotics, and engineering with the use of coding tools.

The lesson design started from a design-based learning approach, viewing the design process as the core project guiding learners' thinking and project completion experiences (Chen & Terken, 2022; Yüksel, 2025). As such, it is a goal-oriented effort to solve tasks

DOI: 10.4324/9781003530343-4

aiming to create a product (Friedman, 2003), which is critical in STEM education. The design presenter here, therefore, integrates hands-on learning and collaborative problem-solving through online teacher professional development that can be applied in all K-12 educational settings. Based on the features of design-based learning found in the domains of project design, teacher roles, assessment, and social context (Gómez Puente et al., 2013), we first designed stages of learning that address each domain as a process. Then, specific features of design-based learning were transformed into four modes of engagement that apply effectively in online learning.

The stages of learning and the four modes of engagement suggest what online CPD providers and instructors should consider ensuring effective content delivery with meaningful engagement among teachers. Drawing from our experience implementing this approach with in-service teachers, we provide practical guidelines for structuring online learning experiences that develop both technical competency and pedagogical understanding. For this chapter, both the stages of learning and the four modes of engagement are presented in the context of an arts integrated robotics/programming/engineering project.

Theoretical Foundations

Our approach builds on design-based learning that is grounded in the processes of inquiry and reasoning for creating solutions, artifacts, or systems in real-life contexts (Gómez Puente et al., 2013; Matere et al., 2023). The broadness of this approach overlaps with the constructionist approach (Papert, 1993), which builds on the constructivist theories of Jean Piaget (Harel & Papert, 1991), highlighting the learners' role as agents who build and construct their own knowledge. This theoretical foundation, derived from the creation of knowledge and the creation of artifacts, is often referred to as maker education in the STEAM context. By engaging in complex projects, maker education enables learners to go through creative making processes by promoting design thinking processes.

While highlighting the importance of producing or engaging in designing activities as a means of learning, the design process offers a valuable learning environment (Matere et al., 2023). When a maker project is designed carefully using the design-based learning approach, learners not only build something tangible but also share work, reflect on it, and refine the products. While the benefits of the process of engaging in design thinking, making processes, and construction of knowledge are evident (Kolodner et al., 2003; Veldhuis et al., 2021), its application in teacher professional development, particularly in online environments, represents an emerging area of investigation. Design-based learning has long been used in design-related curricula in higher education, such as engineering, computer science, and architecture (Matere et al., 2023). This chapter introduces a model involving stages of learning and modes of engagement in a different arena – within the context of arts-integrated STEM education projects.

Literature Review

Arts Integration in STEM Education

The integration of arts into STEM education – creating "STEAM" – has emerged as a significant educational movement aimed at fostering creativity, critical thinking, and innovation among students (Kim & Park, 2012). This approach recognizes that the arts component serves multiple functions: It enhances creativity, promotes innovative problem-solving, and provides multiple pathways for expression and communication (Kim et al., 2023; Kim & Park, 2012). In teacher professional development contexts, arts integration helps educators develop more engaging and inclusive instructional strategies that in turn, can reach diverse learners.

The learning activities introduced in this approach were intentionally designed to combine music with programming, robotics, and engineering with the use of coding tools. Such integration creates opportunities for teachers to experience firsthand the creative processes they will facilitate with their students, promoting creative, interdisciplinary, and real-world problem-solving by adding the arts into STEM education. More importantly, the shared goal inherent in collaborative arts-integrated projects addresses one of the primary challenges of online professional development: isolation and disconnection.

When teachers collaborate toward a common artistic-technical outcome, such as a group performance or collective artwork, they experience the social dimensions of learning that are often missing in online environments. Working toward a shared team goal in arts-integrated projects fosters participants' connectedness, positive emotions, and satisfaction in an online learning environment (An et al., 2022). This collaborative framework not only enhances the learning experience but also models the type of inclusive, team-based pedagogical approaches that teachers can implement in their own classrooms, demonstrating how individual contributions can culminate in meaningful collective achievements. In short, the combination of maker education with arts integration provides authentic learning experiences that mirror the multidisciplinary nature of real-world problem-solving.

From Isolation to Collaboration: Creating Interactive Online Professional Development

The shift to online learning environments has created both challenges and opportunities for professional development (Lay et al., 2020). Traditional barriers such as limited communication, reduced interaction, and feelings of disconnectedness are well documented among online learning communities (Adedoyin & Soykan, 2023; Greenhow et al., 2022). However, online environments also offer unique affordances for flexible, accessible, and scalable professional development experiences. The approach presented here is based on design-based learning, especially with the project sharing a goal together with peers to overcome issues related to isolation. When learners work toward a common objective, the learning experience transcends individual skill development to encompass social, emotional, and community-building dimensions (An et al., 2022). This collaborative framework is particularly unique in the design of our project.

Finally, this collaborative arts-integrated project is taught based on four learning stages, thereby ensuring four different modes of engagement. By intentionally designing for both individual contribution and collective achievement, our approach transforms the potential isolation of online learning into an opportunity for meaningful collaboration. The framework we present in the following section – comprising specific learning stages and modes of engagement – provides a structured, yet flexible approach to implementing hands-on STEAM education in virtual environments, demonstrating authentic maker experiences.

The TMSR (Thinkering, Making, Sharing, Reflection) Model

The development of our TMSR (Thinkering, Making, Sharing, Reflection) Model (Figure 3.1) has emerged from the intersection of constructionist learning theory, design-based learning principles, and the specific requirements of online STEAM professional development. Our model is an adaptation of Kang and Kim's (2017) maker education model TMSI (Tinkering, Making, Sharing, Improving), which was based on a study conducted in an elementary science class. Our TMSR Model (An & Sung, 2024) addresses key features of design-based learning and effective elements needed for online maker projects, for which we replaced Kang and Kim's "Tinkering" with "Thinkering" and "Improving" with "Reflecting."

As a result, in our TMSR Model (see Figure 3.1), "T" stands for "Thinkering," meaning thinking through tinkering. That is, "Thinkering" emphasizes thinking processes while learners are tinkering with their hands (Michalko, 2011; Poce et al., 2019), which is the stage of exploring concepts while experimenting with tools and ideas using one's hands. This phase draws on the constructionist principle that understanding emerges through active manipulation and exploration of materials and ideas (Papert, 1980).

"Making" encompasses the creation phase, where learners build tangible artifacts while simultaneously constructing conceptual understanding (Valente & Blikstein, 2019). This is aligned with design-based learning's emphasis on authentic creation and production as vehicles for learning (Gómez Puente et al., 2013; Veldhuis et al., 2021).

"S" stands for "Sharing," allowing community interaction and collaborative meaning-making. In online environments, deeper sharing can be challenging; however, the current project carefully and intentionally designed shareable goals that learners can achieve together in online environments. Thus, this structured sharing highlights the "idea that happens especially felicitously when the learner is engaged in the construction of something external or at least sharable" (Papert, 1991, p. 3).

The last component, "R," stands for "Reflection," which encompasses individual reflection on learning processes as well as collective reflection on teaching implications. Multiple levels of reflection that lead to the discussion of teaching implications are particularly important and unique in professional development contexts because teachers' learning should be extended to how they will apply this learning in their classrooms.

The TMSR Model provides a comprehensive framework for structuring interactive hands-on projects integrating arts in STEM in online environments. Each stage plays a crucial role in supporting teacher learning and development, leading to a cyclical process of learning and development. Further, each stage builds upon the others, creating a seamless flow between thinking, making, sharing, and reflecting that supports both individual growth and community building. Understanding how these components interact is crucial for designing effective learning experiences.

Table 3.1 presents the key elements of each TMSR stage, specific strategies for online delivery, and their implementation in arts-integrated STEM projects, demonstrating how traditionally hands-on maker activities can be effectively facilitated in virtual environments (Figure 3.2).

Consistent with the definition of arts integration as "an interdisciplinary teaching practice through which non-arts and arts content is taught and assessed equitably in order to

TABLE 3.1 TMSR Model and Its Key Elements, Implementation in Arts-Integrated STEM Projects, and Online Delivery Strategies

Stages	*Key Elements*	*Implementation in Arts-Integrated STEAM Projects*	*Online Delivery Strategies*
Thinkering	• Exploring tools and technologies through guided discovery • Developing an intuitive understanding through hands-on experimentation • Building connections between physical manipulation and conceptual understanding	• Exploring musical concepts while learning coding basics • Experimenting with Makey Makey via guided discovery • Using Scratch coding tutorials in shared screens • Discovering digital-artistic connections together	• Synchronous guided exploration sessions • Screen-sharing for tool demonstrations • Breakout rooms for paired experimentation • Interactive polling for concept checking
Making	• Creating physical and digital artifacts • Integrating arts and technology • Applying design thinking principles • Developing technical skills through authentic projects	• Constructing paper instruments at home • Coding musical pieces in Scratch online, following the steps • Connecting Makey Makey in real-time sessions • Creating individual components for collective work	• Individual creation with periodic check-ins • Document cameras for physical making • Cloud-based coding platforms • Asynchronous work with synchronous support
Sharing	• Present their work to peers • Provide and receive feedback • Develop collaborative problem-solving strategies • Build confidence in their technical and pedagogical skills	• Virtual orchestra performances on Zoom • Sharing instrument designs via camera • Collaborating in breakout rooms • Achieving a collective musical performance online	• Synchronous performance sessions • Gallery walks via screen sharing • Breakout rooms for collaboration • Recording capabilities for documentation
Reflecting	• Individual reflection on learning processes • Collective reflection on teaching implications • Documentation of learning through portfolios • Planning for classroom implementation	• Reflective essays in online forums • Group discussions on pedagogical applications • Digital portfolios documenting the process • Shared planning documents for classroom use	• Individual reflection on learning • Collective teaching implications • Documentation of journey • Implementation planning

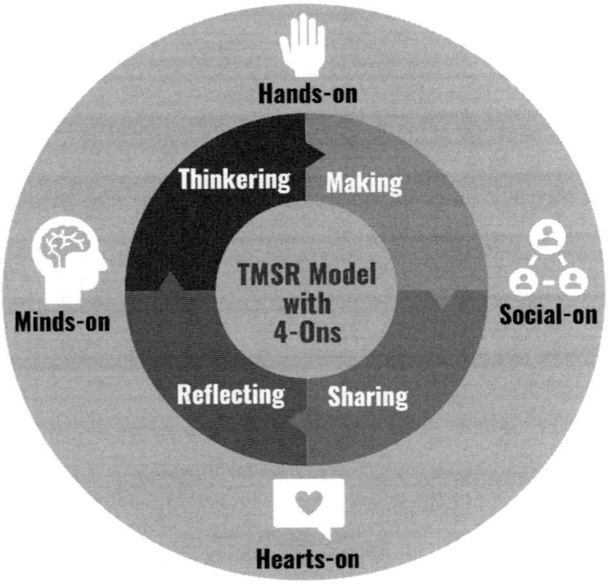

FIGURE 3.1 TMSR Model

deepen students' understanding of both" (New Jersey Principals and Supervisors Association, 2018, p. 9), each stage of the TMSR Model ensures that the arts serve as an integral component rather than a superficial add-on. The framework emphasizes utilizing physical and digital tools while maintaining its collective mindset of playful and collaborative activities (Martin, 2015). Through thoughtful selection of accessible materials, clear guidance for independent work, and strategic use of synchronous and asynchronous collaboration, traditionally hands-on maker activities are effectively adapted for virtual environments.

The implementation column illustrates how music is integrated in STEM contexts, as the online orchestra is completed (An et al., 2022). From exploring musical concepts while learning coding basics in the Thinkering stage to achieving collective orchestra performances in the Sharing stage, each activity builds both technical competency and pedagogical understanding.

The online delivery strategies outlined in Table 3.1 address the critical need to create authentic tasks characterized by high levels of social presence and collaboration, as well as projects involving shared goals and values (Cullen et al., 2013; Holmes, 2013). During synchronous sessions, breakout rooms facilitate small-group work in the Thinkering and Sharing stages, while screen-sharing and document cameras enable real-time demonstrations and presentations. The making stage balances individual creation with periodic check-ins, allowing teachers to work independently while maintaining a connection to the learning community.

This structured approach to online collaboration has been shown to make classes more enjoyable and emotionally fulfilling, with increased participation and affective engagement leading to a greater probability of STEM success (Overland, 2013; Riley, 2012). This comprehensive approach transforms potential isolation in online learning into meaningful collaborative experiences that model the pedagogical strategies teachers can implement with their own students.

Four Modes of Engagement Throughout the TMSR Stages

The effectiveness of maker education in online environments depends heavily on engaging teachers through multiple learning modes simultaneously. While traditional online professional development often focuses primarily on cognitive engagement, our research demonstrates that diverse modes of engagement occur when CPD promotes collaborative, tangible, and reflective learning experiences (An et al., 2022). As such, we identified four essential modes of engagement – hands-on, minds-on, hearts-on, and social-on – that operate both independently and synergistically throughout the TMSR stages (Figure 3.2).

Hands-on engagement involves physical making and doing, where learners actively construct, design, manipulate, draw, examine, and play outcomes. In online environments, this mode requires careful adaptation for authentic maker experiences. For example, teachers engage in hands-on learning by creating paper instruments, connecting Makey Makey circuits, and physically interacting with materials, even while participating remotely.

Minds-on engagement encompasses the cognitive dimensions of learning, where participants actively think about their learning and experiences. This includes understanding coding logic, grasping musical concepts, developing computational thinking, and making connections between arts and technology. Throughout the TMSR stages, teachers continuously engage their minds in problem-solving, debugging code, and translating creative visions into technical implementations.

Hearts-on engagement addresses the emotional and motivational dimensions of learning. Learners engage positively with the project, experiencing a sense of comfort and joy

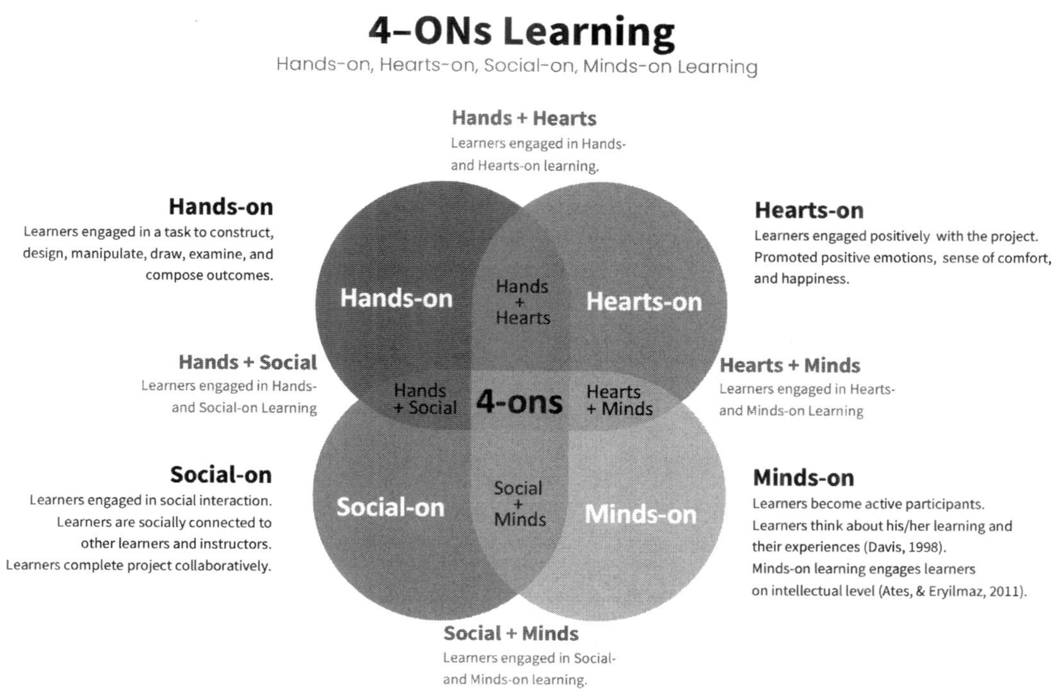

FIGURE 3.2 Four modes of engagement in learning: 4-Ons' descriptions and how each mode can be combined

in creating music through code and playing orchestra virtually together. This emotional engagement is particularly crucial in online environments where isolation can diminish motivation. When teachers experience joy in creating artistic creations and excitement about collaborative performances, they develop the emotional connection necessary for sustained learning.

Social-on engagement involves collaborative interaction and community building. This mode directly addresses one of the primary components of the arts-integrated project, preparing and playing the virtual orchestra together. Through breakout room collaborations, collective orchestra performances, and peer feedback sessions, teachers experience the power of shared goals and collective achievement that An et al. (2022) identified as crucial for successful online maker education.

Combined Modes of Engagement

As illustrated in Figure 3.1, the strength of this framework lies in the intersections where multiple modes combine. For example, when hands-on and social-on merge, learners engage in collaborative making where individual contributions build toward shared goals. Further, the combination of hearts-on and minds-on creates emotionally engaged intellectual exploration, where positive feelings enhance cognitive processing and retention. When hands-on meets hearts-on, learners connect learning to real-life experiences promoting motivation and their interests. Finally, the 4-on engagement created comfortable and positive emotions along with increased confidence leading to the meaningful collaborative learning where everyone's ideas, thoughts, and opinions are welcomed, shared, and respected (An et al., 2022).

The center of the model, where all four modes converge, represents the optimal learning experience. Indeed, this is what An et al. (2022) observed when teachers performed their virtual orchestra. In these moments, participants were simultaneously creating (hands-on), problem-solving (minds-on), experiencing positive emotions (hearts-on), and collaborating toward a shared achievement (social-on).

Throughout the TMSR stages, these modes of engagement occurred differently at different times and occasions. For example, during Thinkering, hands-on and minds-on engagement mainly played a role as teachers explored tools and concepts. Making involved hands-on engagement while maintaining cognitive challenge. The Sharing stage promoted social-on engagement. The hearts-on stage was also engaged while participants were collectively preparing for the final goal and playing together. Finally, reflection engaged minds-on, social-on, and hearts-on together. Understanding these patterns enables CPD facilitators to design activities that maintain multi-modal engagement throughout the learning journey.

Examples of Applying the TMSR Model

We applied the TMSR Model to the design of an online module for in-service teachers in a graduate class and evaluated its effectiveness. A detailed description of this application is available in An et al. (2022). The following is an additional example of how the TMSR Model can be used to integrate the arts into STEM instruction. Table 3.2 presents the step-by-step activities organized according to the TMSR Model.

The project shown in this blog page: https://tinyurl.com/mpe5fa8m, was developed around the theme of "unity through diversity," in which each student contributed a tile

TABLE 3.2 Example: Activities Organized by the TMSR Model

Thinkering	• Introduce the theme of "unity through diversity" and show examples of grid-based art. • Have students brainstorm and sketch possible tile designs using geometric shapes, handprints, or symbolic imagery. • Explore mathematical ideas such as proportion, symmetry, and scale while experimenting with layout options.
Making	• Each student creates a 6″ × 6″ tile using clay, cardboard, or paper. • Students decorate tiles with patterns, colors, or symbols that represent their identity. • The teacher incorporates math skills (measuring dimensions, calculating total area) and art techniques (color mixing, glazing, or shading).
Sharing	• Combine individual tiles to form a collective wall installation. • Host a gallery walk or digital showcase where students explain their design choices. • Discuss how individual contributions connect to the shared theme.
Reflection	• Students write or record short reflections on what they learned about geometry, art, and community through the project. • Teachers guide a discussion on how this process connects to broader STEM concepts (e.g., tessellation, patterns) and classroom collaboration.

by imprinting their handshape and including symbolic imagery representing their cultural background. The tiles, made from clay, cardboard, or paper, were then assembled into a large heart-shaped mural (other shapes could also be chosen). The final outcome, along with photos documenting the step-by-step process, is available on the blog page.

Conclusion and Implications

This chapter has demonstrated how arts-integrated maker projects can transform online teacher professional development by addressing the challenges of isolation, disconnection, and limited hands-on engagement often found in online contexts. The TMSR framework and four modes of engagement provide a structured approach that enables CPD providers to create meaningful, collaborative learning experiences in virtual environments. The success of this approach, as evidenced by increased teacher satisfaction, enhanced community building, and engagement (An et al., 2022), offers important implications for practice.

For effective implementation, CPD providers need to consider both pedagogical and practical components. Online platforms, technological tools, and hands-on manipulatives need to be accessible for all online learners. More importantly, collaboration opportunities need to be intentionally and carefully designed through structured interactions and collective achievements.

As online professional development continues to evolve, the integration of arts, making, and collaborative learning offers a promising path forward. The approach presented in this chapter not only develops technical competencies but also nurtures the creativity, community, and pedagogical innovation that are essential for 21st-century STEM education. Through careful attention to the stages of learning and modes of engagement outlined in this chapter, CPD providers can transform online professional development from an isolated, cognitive exercise to a more engaging experience that prepares teachers for STEAM education.

KEY TAKEAWAYS

- The TMSR Model transforms online isolation into collaboration by structuring learning through four sequential stages: Thinkering (hands-on exploration), Making (artifact creation), Sharing (collaborative presentation), and Reflection (individual and collective analysis), thereby creating a cyclical process that builds both technical skills and pedagogical understanding.
- The model's four modes of engagement work synergistically to maximize learning: hands-on (physical making), minds-on (cognitive processing), hearts-on (emotional connection), and social-on (collaborative interaction), with effective learning occurring when all four modes converge simultaneously.
- Arts integration serves as a powerful catalyst for overcoming traditional online learning challenges by providing shared creative goals that foster community building, enhance engagement, and model interdisciplinary problem-solving approaches that teachers can implement in their own classrooms.
- Collaborative maker projects with shared outcomes address the disconnection commonly experienced in virtual environments by transforming individual contributions into collective achievements, such as virtual orchestra performances that combine individual coded musical pieces into group performances.
- Strategic use of synchronous and asynchronous activities enables authentic hands-on experiences in online settings through carefully designed combinations of breakout rooms, screen-sharing, document cameras, and individual creation periods with periodic check-ins.
- The effectiveness of professional development increases significantly when teachers experience the pedagogical approaches firsthand, allowing them to understand both the technical implementation and the learning processes they will facilitate with their own students.

References

Adedoyin, O. B., & Soykan, E. (2023). Covid-19 pandemic and online learning: The challenges and opportunities. *Interactive Learning Environments*, *31*(2), 863–875. https://doi.org/10.1080/10494820.2020.1813180

An, H., & Sung, W. (2024). The thinkering, making, sharing, and reflecting (TMSR) model: An instructional design model for online collaborative project-based learning. *Educause Review*. https://er.educause.edu/articles/2024/2/thinkering-making-sharing-and-reflecting-a-model-for-online-collaborative-project-based-learning

An, H., Sung, W., & Yoon, S.Y. (2022). Hands-on, minds-on, hearts-on, social-on: A collaborative maker project integrating arts in a synchronous online environment for teachers. *TechTrends*, *66*(4), 590–606. https://doi.org/10.1007/s11528-022-00740-x

Angelino, L. M., Williams, F. K., & Natvig, D. (2007). Strategies to engage online students and reduce attrition rates. *Journal of Educators Online*, *4*(2), n2. Retrieved March 28, 2022, from http://www.thejeo.com/Volume4Number2/Angelino%20Final.pdf

Ateş, Ö., & Eryilmaz, A. (2011). Effectiveness of hands-on and minds-on activities on students' achievement and attitudes towards physics. *Asia-Pacific Forum on Science Learning and Teaching*, *12*(1), 1–22. https://go.openathens.net/redirector/uttyler.edu?url=https://www.proquest.com/scholarly-journals/effectiveness-hands-on-minds-activities-students/docview/1955058563/se-2

Bolliger, D. U., & Inan, F. A. (2012). Development and validation of the online student connectedness survey (OSCS). *International Review of Research in Open and Distributed Learning*, *13*(3), 41–65.

Bragg, L. A., Walsh, C., & Heyeres, M. (2021). Successful design and delivery of online professional development for teachers: A systematic review of the literature. *Computers & Education*, *166*, 104158.

Chen, F., & Terken, J. (2022). *Design process*. In *Automotive interaction design: From theory to practice* (pp. 165–179). Springer Nature Singapore.

Chiu, T. K., Lin, T. J., & Lonka, K. (2021). Motivating online learning: The challenges of COVID-19 and beyond. *The Asia-Pacific Education Researcher*, *30*(3), 187–190.

Cullen, R., Kullman, J., & Wild, C. (2013). Online collaborative learning on an ESL teacher education programme. *ELT Journal*, *67*(4), 425–434.

Davis, O. L., Jr. (1998). Beyond beginnings: From "hands-on" to "minds-on". *Journal of Curriculum and Supervision*, *13*(2), 119–136. https://go.openathens.net/redirector/uttyler.edu?url=https://www.proquest.com/scholarly-journals/beyond-beginnings-hands-on-minds/docview/196370071/se-2

Friedman, K. (2003). Theory construction in design research: Criteria: Approaches, and methods. *Design Studies*, *24*(6), 507–522. https://doi.org/10.1016/S0142-694X(03)00039-5

Gómez Puente, S. M., Van Eijck, M., & Jochems, W. (2013). A sampled literature review of design-based learning approaches: A search for key characteristics. *International Journal of Technology and Design Education*, *23*(3), 717–732. https://doi.org/10.1007/s10798-012-9212-x

Greenhow, C., Graham, C. R., & Koehler, M. J. (2022). Foundations of online learning: Challenges and opportunities. *Educational Psychologist*, *57*(3), 131–147. https://doi.org/10.1080/00461520.2022.2090364

Harel, I., & Papert, S. (Eds.). (1991). *Constructionism*. Ablex Publishing.

Holmes, B. (2013). School teachers' continuous professional development in an online learning community: Lessons from a case study of an eTwinning learning event. *European Journal of Education*, *48*(1), 97–112.

Kang, I., & Kim, H. (2017). Exploring the value of the maker mind set at maker education. *The Journal of the Korea Contents Association*, *17*(10), 250–267.

Kim, N. P., Park, S., Kim, Y., Kim, B., Song, C., & Song, N. (2023). Creative fusion: Exploring the convergence of art, science, and education in traditional crafts through bird-inspired design techniques. *International Journal of Arts, Sciences and Education*, *4*(3), 14–27. https://www.ijase.org/index.php/ijase/article/view/269

Kim, Y., & Park, N. (2012). Development and application of STEAM teaching model based on the Rube Goldberg's invention. In A. Amine, L. Bellatreche, Z. Elberrichi, E. J. Neuhold, & R. Wrembel (Eds.), *Computer science and its applications* (pp. 693–698). Springer Netherlands.

Kolodner, J. L., Camp, P. J., Crismond, D., Fasse, B., Gray, J., Holbrook, J., Puntambekar, S., & Ryan, M. (2003). Problem-based learning meets case-based reasoning in the middle-school science classroom: Putting learning by Design™ into practice. *Journal of the Learning Sciences*, *12*(4), 495–547.

Lay, C. D., Allman, B., Cutri, R. M., & Kimmons, R. (2020, September). Examining a decade of research in online teacher professional development. *Frontiers in Education*, *5*, 573129.

Lo, C. K. (2021). Design principles for effective teacher professional development in integrated STEM education: A systematic review. *Educational Technology & Society*, *24*(4), 136–152.

Martin, L. (2015). The promise of the maker movement for education. *Journal of Pre-College Engineering Education Research*, *5*, 30–39. https://doi.org/10.7771/2157-9288.1099

Matere, I. M., Weng, C., Astatke, M., Hsia, C. H., & Fan, C. G. (2023). Effect of design-based learning on elementary students computational thinking skills in visual programming maker course. *Interactive Learning Environments*, *31*(6), 3633–3646.

Michalko, M. (2011). *Creative thinkering: Putting your imagination to work*. New World Library.

New Jersey Principals and Supervisors Association (NJPSA). (2018). *New Jersey's arts integration think and do workbook: A practical guide to think about and implement arts integration*. https://www.artsednj.org/wp-content/uploads/artsintegrationWorkbook2018-1.pdf

Overland, C. T. (2013). Integrated arts teaching: What does it mean for music education? *Music Educators Journal*, *100*(2), 31–37.

Papert, S. (1980). *Mindstorms. children, computers, and powerful Ideas*. Basic Books.

Papert, S. (1991). Situating construction. In I. Harel & S. Papert (Eds.), *Constructionism* (pp. 1–12). Ablex Publishing.

Papert, S. (1993). *The children's machine: Rethinking school in the age of the computer*. Basic Books.

Poce, A., Amenduni, F., & De Medio, C. (2019). From tinkering to thinkering. Tinkering as critical and creative thinking enhancer. *Journal of e-Learning and Knowledge Society*, *15*(2). https://doi.org/10.20368/1971-8829/1639
Quigley, C. F. (2016). "Finding the joy in the unknown": Implementation of STEAM teaching practices in middle school science and math classrooms. *Journal of Science Education and Technology*, 1–17. https://doi.org/10.1007/s10956-016-9602-z
Riley, S. (2012). *STEAMpoint: A guide to integrating science, technology, engineering, the arts, and mathematics through the common core*. Education Closet
Taysever, G. J. (2016). *The effects of online professional development on teacher behavior and perceptions of science, technology, engineering, art and math teaching efficacy* [Doctoral dissertation, Governors State University]. https://opus.govst.edu/capstones/233
Valente, J. A., & Blikstein, P. (2019). Maker education: Where is the knowledge construction? *Constructivist Foundations*, *14*(3), 252–262. https://constructivist.info/14/3/252
Veldhuis, A., d'Anjou, B., Bekker, T., Garefi, I., Digkoglou, P., Safouri, G., Safouri, G., Remotti, S., Beamer Cronin, E., & Bouros, M. (2021, June). The connected qualities of design thinking and maker education practices in early education: A narrative review. In *FabLearn Europe/MAKEED 2021 – An international conference on computing, design and making in education* (pp. 1–10). Association for Computing Machinery. https://doi.org/10.1145/3466725.3466729
Wong, J. T., Bui, N. N., Fields, D. T., & Hughes, B. S. (2022). A learning experience design approach to online professional development for teaching science through the arts: Evaluation of teacher content knowledge, self-efficacy and STEAM perceptions. *Journal of Science Teacher Education*, *34*(6), 593–623. https://doi.org/10.1080/1046560X.2022.2112552
Yüksel, A. O. (2025). Design-based STEM activities in teacher education and its effect on pre-service science teachers' design thinking skills. *Journal of Science Education Technology*, *34*, 904–918. https://doi.org/10.1007/s10956-025-10215-2

4

ONLINE PROFESSIONAL DEVELOPMENT ON INTEGRATING LITERATURE INTO STEM EDUCATION TO CREATE MEANINGFUL CONNECTIONS

The ATTACK Model

Gloria Campbell-Whatley and Kimberly Bunch-Crump

Introduction

In the ever-evolving field of science, technology, engineering, and mathematics (STEM) education, it is essential to foster meaningful and productive learning experiences for all students. However, many teachers find it challenging to make STEM content accessible, engaging, and relevant for their students while maintaining rigorous standards. Integrating literature into STEM education offers a powerful and promising pathway to bridge learning challenges by providing contextualized, human-centered entry points that support culturally responsive teaching.

Effective STEM education requires not only mastery of content knowledge but also critical thinking, problem-solving, and meaningful connections to students' lives and experiences. Literature provides a meaningful way to introduce STEM concepts through real-world contexts, aligned with constructivist and inquiry-based learning approaches. Integrating literature into STEM instruction fosters effective instruction by providing narratives and contexts that humanize STEM content and highlight its real-world applications. These approaches emphasize building on the learner's prior knowledge and experiences while promoting exploration, dialogue, and reflection.

This chapter provides a rationale for integrating literature into STEM and provides a framework with practical strategies for training in-service teachers in online professional development (PD) formats to create effective STEM learning environments supportive of all students.

Integrating STEM and Literature into Online PD

As they attempt to balance providing flexible and economical options to equip in-service educators with effective strategies for adapting curricula, maintaining engagement and interactions, providing hands-on experiences, and managing resources and materials, schools routinely turn to continuous PD in online settings. Effective continuous STEM PD offers integrated solutions that combine online education with practical experiences and innovative

DOI: 10.4324/9781003530343-5

teaching approaches to improve knowledge of STEM concepts and effective teaching strategies (Joseph & Uzondu, 2024; Saldívar-Almorejo et al., 2024).

Integrating literature with STEM ideals presents a novel and dynamic approach to online PD for STEM educators. Indeed, recent studies have demonstrated that integrating STEM and literature can significantly enhance both teaching strategies and, ultimately, student learning outcomes (Boon, 2019; English, 2019; English & King, 2019; McKinney et al., 2017).

Challenges of Integrating Literature in STEM Online PD

While online continuous PD offers flexibility and accessibility for participants, integrating STEM and literature using this format can present challenges. That is, in addition to the expected challenges – shifts in teacher mindset from traditional textbook-driven instruction to incorporating narrative-based instruction, along with time constraints due to competing responsibilities – implementing PD in online formats can present challenges related to technology, sustained engagement, collaborative opportunities, and resource accessibility.

Technological barriers, such as teacher unfamiliarity with interactive digital platforms requiring technical support from continuous PD facilitators, can lead to frustration and training disruptions. Additionally, technology may reduce the accessibility of literature for collaborative online activities due to copyright or restrictions in the learning management system (LMS; e.g., Canvas, Blackboard). Online STEM PD can also create engagement and participation challenges (Saldívar-Almorejo et al., 2024), such as screen fatigue, incomplete or inconsistent engagement with materials, and limited opportunities for face-to-face discussions that foster deeper understanding.

Further, challenges related to collaboration can hinder the in-service teacher's development of practical strategies for classroom implementation. Additionally, resource constraints (e.g., accessibility to literature that aligns with STEM standards, hands-on activities related to literature, and physical access to materials and manipulatives) and the pedagogical challenges of designing literature-integrated STEM instruction in a digital environment require intentional support to ensure effective teacher learning. Despite these obstacles, however, integrating literature into STEM-focused PD offers powerful benefits that can transform teaching practice and enrich student learning given thoughtful design and support.

Benefits of Integrating Literature in Online STEM PD

Integrating literature into STEM continuous PD offers PD facilitators and participating in-service teachers a powerful tool for deepening content knowledge, enhancing instructional practices, and making STEM learning more accessible and engaging for all learners. By situating STEM concepts within meaningful narratives and real-world contexts, literature supports the development of culturally responsive pedagogy and fosters connections to participants' lived experiences. Such an approach not only reinforces constructivism and inquiry-based learning but also provides educators with concrete strategies for promoting critical thinking, communication, and engagement across disciplines.

A plethora of pedagogies that advance STEM teaching and learning have been developed, investigated, and validated. For example, an examination of 229 published articles focused on instructional methodologies identified design-based learning, inquiry-based learning, project-based learning, and problem-based learning as crucial approaches to equipping

21st-century educators with the skills necessary to thrive in the rapidly evolving STEM fields (Halawa et al., 2024), with a special emphasis on the development of critical analytical and logical thinking skills (Grangeat et al., 2021; Sahito & Wassan, 2024; Van Hoe et al., 2024).

Inquiry- and Problem-Based Learning

Inquiry- and problem-based learning are particularly aligned with the scientific process, as they encourage investigation through problem-solving questions or scenarios instead of simply presenting established facts. Additionally, inquiry-based learning complements problem-based learning methods, which involve research exploration aimed at developing practical, logical, and relevant solutions. Together, these methodologies can motivate and maintain learner engagement when aligned with educators' professional interests and relevant, interactive content. For example, they provide opportunities for collaborative learning and sharing practices, thereby helping to foster a sense of community. Additionally, structured reflection on both content and pedagogy helps educators integrate new knowledge into their teaching.

Specifically with regard to integrating literature into STEM instruction, this approach supports inquiry- and problem-based methodologies and offers numerous benefits, including enhanced comprehension and engagement, development of critical thinking and problem-solving skills, fostering of empathy and diversity of perspective, and increased accessibility – all of which contribute to a strong foundation for learning (Hadaway et al., 2018; Mahzoon-Hagheghi et al., 2018; Pantaleo, 2020; Rice, 2012; Sahito & Wassan, 2024).

Literature contextualizes scientific concepts, making them more accessible, relatable, and memorable, while also increasing learner interest through narrative-driven content (Pantaleo, 2020). By humanizing STEM knowledge and presenting authentic, real-world problems, literature fosters inquiry-based learning and deeper engagement. Research further supports that literature-integrated STEM instruction not only strengthens literacy skills but also improves the learner's interest and performance in science (Sahito & Wassan, 2024).

Literature-integrated STEM continuous PD activities engage in-service teachers in narrative-based challenges like the inquiry and problem-solving processes foundational to effective STEM instruction (Hadaway et al., 2018). Literary works often present problems requiring logical reasoning and creative solutions, providing models that teachers can use in their classrooms. In addition, discussing and writing about literature within continuous PD enhances teachers' communication skills, specifically articulating thoughts and instructional strategies in writing, resulting in enhanced clarity in lesson delivery and facilitation.

By connecting scientific concepts to human experiences, the integration of literature in STEM continuous PD broadens teachers' understanding of STEM while fostering empathy and awareness of societal and ethical dimensions (Mahzoon-Hagheghi et al., 2018; Rice, 2012). Engaging with diverse viewpoints helps educators integrate sociopolitical themes into instruction, encouraging critical thinking and inclusivity. Literature demystifies STEM by making it more relatable and accessible, inspiring teachers to design interdisciplinary, student-centered lessons (Pantaleo, 2020). This approach nurtures a holistic teaching mindset and equips educators with tools to foster curiosity and meaningful learning across cultures, an outcome that is supported by the National Science Foundation's push for diversity in STEM to combat racial and gender disproportionality in the field (Wright, 2018).

Indeed, significant racial, ethnic, and gender disparities persist in both STEM education and the STEM teaching workforce. For example, Black, Hispanic, and Indigenous students constitute approximately 35% of students planning to pursue STEM in college but earn only about 23% of STEM degrees (Pew Research Center, 2021a). Further, women, although earning a majority of college degrees nationwide, are notably underrepresented in several STEM fields, particularly engineering and computer science (Pew Research Center, 2021a).

This lack of diversity extends to the STEM teaching profession. While minority students constitute nearly 60% of public-school K-12 students, only about 20%–25% of STEM teachers come from minority backgrounds (Pew Research Center, 2021b; UNESCO GEM Report, 2024). This mismatch limits culturally affirming learning experiences for many students and underscores the urgent need for cultural competence and responsive teaching in STEM.

Culturally Responsive Teaching

While it emphasizes inquiry and integration, STEM education often underemphasizes culturally responsive practices. Without PD with an intentional focus on inclusiveness, these gaps will persist. Integrating literature into STEM continuous PD offers a powerful pathway to support learners and build a more inclusive, engaging STEM experience for all.

Culturally responsive teaching in STEM begins with building cultural competence among educators. Therefore, continuous PD should guide teachers in reflecting on their identities, recognizing diverse student experiences, and integrating culturally relevant and sustaining pedagogies into their instruction (Alim et al., 2020; Bunch-Crump & Hitt, 2023; Milner, 2012). These approaches, grounded in academic achievement, sociopolitical awareness, and cultural inclusion, support learning environments that validate student voice and identity (Johnson & Bojorquez, 2022; Ladson-Billings, 2021). Specifically, *culturally sustaining pedagogy* emphasizes preserving students' heritage and multilingualism, whereas *culturally responsive pedagogy* uses students' lived experiences to enhance instruction and engagement.

Integrating literature into STEM continuous PD complements these practices by humanizing STEM content, fostering empathy, and strengthening comprehension, creativity, and communication (Hadaway et al., 2018; Pantaleo, 2020). Stories grounded in STEM themes offer accessible entry points for inquiry, ethics, and real-world problem-solving. Educators report challenges in implementing culturally responsive teaching, using student backgrounds as assets in teaching and learning, especially in virtual settings, due to limited training and support (Kim et al., 2024). Literature-based activities can serve as practical tools to promote inclusive dialogue, broaden perspectives, and support interdisciplinary learning (Mahzoon-Hagheghi et al., 2018; Rice, 2012). Continuous PD should provide diverse texts, reflection opportunities, and culturally responsive strategies to support both teacher development and learner engagement (Delpit, 2012; Eden et al., 2024).

To sustain equity-driven STEM education, continuous PD must evolve with the field's growing complexity. High-quality, accessible, and inclusive professional learning is essential, particularly for under-resourced or rural educators (Ogodo, 2022; Santiago, 2019). As STEM industries anticipate significant job growth by 2033 (U.S. Bureau of Labor Statistics, 2024), continuous PD must promote not only technical skills but also ethical and cultural fluency. Integrating literature into STEM instruction equips teachers with tools to support student identity, deepen interdisciplinary learning, and foster long-term interest in STEM careers. Similarly, online preservice teacher programs play a critical role in preparing culturally competent, future-ready STEM educators through scalable, inclusive training (Chapman, 2023; Lott et al., 2013).

The ATTACK Model (Application, Teaching Strategies, Trial/Try-Out, Assessment, Cognitive Reflection, and Keep, Retain, Generalize)

The ATTACK model (Campbell-Whatley et al., 2021) provides a structured approach to lesson planning that helps ensure engagement, accessibility, and retention. As such, it offers a helpful way for facilitators to structure continuous PD while modeling a viable lesson planning process for educators to use in their own instruction. Within the context of this chapter, the framework can be used to develop PD series designed to equip STEM educators with the knowledge, tools, and confidence to integrate literature into their instruction. The following section describes each component of the ATTACK model with concrete examples to guide implementation in both PD settings and K-12 classrooms.

The Elements of the ATTACK Model

The model begins with **Application**, where lesson objectives and standards are clearly outlined. Next, **Teaching Strategies** focus on delivering content effectively through materials, timing, and essential vocabulary. **Trial/Try-Out** encourages hands-on learning with technology integration and scaffolding for diverse learners. **Assessment** measures educators' progress through various evaluation methods. **Cognitive Reflection** fosters metacognition by encouraging educators to analyze their learning processes. Finally, **Keep, Retain, Generalize** helps educators apply knowledge in different contexts, ensuring long-term understanding and cross-disciplinary connections.

Using the ATTACK Model to Design K-12 Literature-Integrated STEM Lessons

The ATTACK model provides a structured and flexible approach for K-12 educators to design effective literature-integrated STEM lessons that engage students in both reading and scientific inquiry. By guiding teachers through its six key phases: Application, Teaching Strategy, Trial/Try-Out, Assessment, Cognitive Reflection, and Keep/Retain/Generalize, the model ensures that lessons are intentionally planned, standards-aligned, and culturally responsive. Each component supports the integration of high-quality literature with STEM content, allowing educators to connect narrative elements to real-world scientific problems, promote student-led inquiry, and deepen understanding through authentic assessment and reflection.

Whether designing a unit on genetics using *The Immortal Life of Henrietta Lacks* (Skloot, 2011) or exploring engineering concepts through biographies of inventors, ATTACK empowers teachers to plan lessons that make STEM more accessible, meaningful, and inclusive for all students. Table 4.1 displays the elements of the ATTACK model when used to plan literature-integrated STEM lessons, including definitions, instructional uses, and planning activities. Table 4.2 provides an example of using the model to plan learning activities focused on a common learning goal.

TABLE 4.1 ATTACK Model for Planning Literature-Integrated STEM Lessons

Component	*Definition*	*Use*	*Activities*
Application	Teachers identify specific learning goals for STEM and literacy that students will achieve through a literature-integrated lesson.	Ensures alignment with standards and addresses student needs.	• Set clear, measurable objectives • Align lessons with national/state STEM and English Language Arts (ELA) standards • Plan real-world connections for STEM concepts
Teaching Strategy	Teachers select high-interest, relevant literature that aligns with STEM concepts and design instructional strategies.	Introduces STEM ideas through storytelling and inquiry-based methods.	• Choose diverse, STEM-aligned texts • Plan discussion questions linking the story to STEM • Design inquiry and exploration activities
Trial/ Try-Out	Teachers rehearse and implement lessons in low-stakes ways while gathering informal feedback.	Allows lesson refinement based on student engagement and peer collaboration.	• Conduct mini-lessons or segments • Gather student feedback • Collaborate with colleagues for refinement
Assessment	Teachers create tools to measure student comprehension of STEM content and connections to the story.	Evaluates the effectiveness of the lesson and student learning outcomes.	• Develop rubrics and checklists • Use exit tickets or student journals • Review student work for comprehension and STEM connections
Cognitive Reflection	Teachers reflect on student engagement, goal attainment, and areas for improvement or differentiation.	Guides future lesson adjustments and instructional strategies.	• Use journaling and discussions within professional learning communities to reflect on lessons • Analyze student engagement and outcomes • Adjust instructional methods as needed
Keep/Retain/ Generalize	Teachers adapt and repeat the lesson with different texts or STEM topics, collaborating and embedding strategies across the curriculum.	Promotes sustainability and scalability of literature-integrated STEM practices.	• Adapt lessons for new contexts • Share practices with peers • Embed strategies into interdisciplinary units

TABLE 4.2 Literature-Integrated STEM Lesson Using ATTACK

<table>
<tr><td colspan="2">Biology Lesson Plan
Grade Level: 9–10
Implementation Timeline: Weeklong mini unit
STEM Standard(s):
• NGSS HS-LS1-1: Construct an explanation based on evidence for how the structure of DNA determines the structure of proteins.
• NGSS HS-ETS1-3: Evaluate a solution to a complex real-world problem based on prioritized criteria and trade-offs.
Literacy Standard(s):
• CCSS.ELA-LITERACY.RI.9-10.2: Determine a central idea of a text and analyze its development.</td></tr>
<tr><td>ATTACK Component</td><td>Examples of Activity Options for Various Lessons</td></tr>
<tr><td>A – Application
Begin by reviewing and selecting relevant standards.
• Unpack the standard, identifying the content knowledge, skills, concepts/understanding, and vocabulary necessary to meet the standard.</td><td>Objectives:
STEM Standard(s):
• NGSS HS-LS1-1: Construct an explanation based on evidence for how the structure of DNA determines the structure of proteins.
• NGSS HS-ETS1-3: Evaluate a solution to a complex real-world problem based on prioritized criteria and trade-offs.
Literacy Standard(s):
CCSS.ELA-LITERACY.RI.9-10.2: Determine a central idea of a text and analyze its development.</td></tr>
<tr><td>• Create a broad learning goal – an outcome achieved by meeting 3–5 daily outcome objectives.</td><td>Learning Goal: Students will understand how HeLa cells contributed to scientific discovery while critically evaluating ethical research practices.</td></tr>
<tr><td>• Create lesson objectives.</td><td>Lesson Objective (observable and measurable): Students will summarize the impact of HeLa cells on science and evaluate the ethical implications of using biological samples without consent.</td></tr>
<tr><td>• Consider how to present the learning objective(s) in student-friendly language.</td><td>"Today you will be able to tell me what a HeLa cell is, how they benefited science, and describe ethical considerations when using biological samples."</td></tr>
<tr><td>T – Teaching Strategy
• Select the text.
  • Identify key excerpts.
  • Identify STEM concepts embedded in text – cell reproduction, cancer cells, ethical considerations in science</td><td>• Text Title and Author: The Immortal Life of Henrietta Lacks by Rebecca Skloot
  • Identify excerpts that describe Henrietta's diagnosis, the extraction of her cells, and their scientific contributions.
• STEM concepts embedded in text: Cell reproduction, cancer cells, ethical considerations in science.</td></tr>
</table>

(Continued)

TABLE 4.2 (Continued)

ATTACK Component	*Examples of Activity Options for Various Lessons*
• Identify additional information/supplemental text.	• Pair excerpts with informational texts and videos about DNA, protein synthesis, HeLa's role in biomedical research, and NGSS's concepts of ethical dilemmas. • Video: *Immortal Cells Turn 96: Henrietta Lacks*

• Consider how to introduce the lesson topic to the students.
  • How will you "launch" the text?
  • How will you use the literary text to make the topic relevant to students?
  • How will you connect the new learning to prior knowledge?
	• Have students read the excerpt from the text and identify the impact of HeLa cells on research and medicine.
	  • Students answer guided questions as they read. What is a HeLa cell? Where did they come from? Why are they important?
	• Use Turn and Talk to discuss: "What is cancer?"
	• Use a whole-group discussion format to discuss: "What is informed consent and how does it apply to science?"

• Identify instructional strategy/approach to introduce and explain concepts and offer students an opportunity to practice with support (e.g., modeling, inquiry)	• Teacher-Led Debate: Teacher provides students with guiding questions to consider both sides of the ethical issue. In pairs, students write and rehearse short pro/con statements using evidence from texts.
	  • Skills: Evaluating ethical implications
	  • Scaffolds: Use of sentence stems and example arguments/claims
	• Jigsaw Text Analysis: Students are assigned different sections of *The Immortal Life of Henrietta Lacks* or related articles. In expert groups, they analyze their section for scientific contributions and ethical issues. Then, they share insights in whole groups.
	  • Skills: Summarizing scientific impact and ethical issues
	  • Scaffolds: Sentence starters, vocabulary word banks, graphic organizers
	• Timeline and Cause-and-Effect Map: In small groups, students co-construct a timeline of key HeLa discoveries and pair them with cause-effect statements (e.g., "HeLa cells were used in the polio vaccine, which enabled mass production and testing.")
	  • Skills: Connecting HeLa events to scientific advancements
	  • Scaffolds: Template, model

• Identify instructional strategy/approach to provide students with the opportunity to practice independently	• HeLa Cell Impact Summary Paragraph: Students write a paragraph summarizing how HeLa cells contributed to scientific discovery, citing at least two specific examples. • Ethical Evaluation Quick Write: Students respond to the prompt: "Was it ethical to use Henrietta Lacks' cells without her consent? Defend your answer with evidence."

(Continued)

TABLE 4.2 (Continued)

ATTACK Component	*Examples of Activity Options for Various Lessons*
	• Central Idea: Students independently annotate a new, short informational article on biomedical ethics, identifying the central idea and its development using evidence from the text.
• Identify instructional strategy/ approach to provide closure and assess student learning	• 3-2-1 Exit Ticket: Students write 3 things they learned about HeLa cells, 2 ethical questions they still have, and 1 way in which this relates to modern science practices. • Think-Pair-Share: "How could what we learned today influence the way you think about medical treatment or consent in real life?" Students share with a partner, then volunteer to share with the whole group.
• Identify plans for relating to personal, cultural, and community relevance	• Group discussion on the intersection of race, class, access to medical care, and informed consent. • Extension Activity: Students research historical cases of medical exploitation and share reflections.
T – Trial/Try-Out • Implement the plan	• Teacher implements the plan and keeps notes on what did/not work and adjustments to future instruction based on the lesson delivery and results.
A – Assessment • Identify how the lesson objective will be assessed. • Document assessment results.	• Claim-Evidence-Reasoning Essay: "Should scientists be allowed to use biological samples from patients without explicit consent?" Use a rubric to evaluate the clarity of the claim, quality of evidence from the text, and strength of reasoning. • HeLa Concept Map: Students create a concept map connecting DNA structure, protein synthesis, HeLa cell replication, and scientific discovery. Use a rubric to evaluate accuracy and depth of connections. • Multiple-choice quiz.
C – Cognitive Reflection • Reflect on student engagement, goal attainment, and areas for improvement or differentiation.	• What connections did students make between science and ethics? • Which parts of the lesson prompted the most engagement? • Which population/group of students was engaged?
K – Knowledge Retention • Identify ways to adapt the lesson for new concepts.	• How can I revise this lesson to incorporate additional concepts, subjects, or content? • How can I use the lesson format to integrate literature into other lessons?

Using the ATTACK Model to Design Literature-Integrated Online STEM Professional Development

The ATTACK model can be used not only to plan K-12 lesson activities but also as a guide to prepare effective PD for educators. Literature-integrated STEM continuous PD should include activities supporting three core practices: (a) collaborative book studies that connect literature to STEM topics; (b) guidance in selecting texts and designing related STEM activities; and (c) opportunities to co-plan interdisciplinary units that integrate literature, STEM inquiry, and hands-on experimentation.

Core Practices of Literature-Integrated STEM PD

Collaborative Book Studies: During collaborative book studies, the PD facilitator can host journal clubs or discussion groups where participating educators read and discuss literature that connects to STEM content. For example, participants can be guided in making thematic connections, identifying opportunities for inquiry-based exploration, and reflecting on ethical or cultural issues (e.g., *The Immortal Life of Henrietta Lacks*).

Selecting Texts and Designing Related STEM Activities: When offering guidance in selecting texts and STEM activity design, facilitators can provide frameworks for choosing culturally relevant and age-appropriate texts that align with math, science, and engineering content. In addition, they should model how to design meaningful, hands-on activities such as experiments, investigations, or real-world problem-solving tasks linked to the stories.

Interdisciplinary Co-Planning: To support interdisciplinary co-planning, facilitators can create structured opportunities for teachers to co-develop interdisciplinary units that integrate STEM concepts, literacy skills, and inquiry-based learning. Educators can work in small groups to design units that include reading, discussion, STEM experimentation, and assessment.

Well-designed continuous PD positively impacts teacher self-efficacy and instructional quality (Zhou et al., 2023). Effective continuous PD also includes sustained learning opportunities, collaborative reflection, and mentoring or coaching support. These elements help teachers internalize instructional strategies and confidently implement them in their classrooms.

The above activities lay a strong foundation for the design and implementation of literature-integrated STEM PD and can be further organized and enhanced using the ATTACK model. By aligning collaborative book studies, text selection guidance, and interdisciplinary co-planning with the core components of the ATTACK model, PD presenters can create a cohesive and practical structure that supports both online professional learning and classroom instruction.

The following section describes each component of the ATTACK model with concrete examples to guide the development of virtual or in-person literature-integrated STEM PD.

Components of the ATTACK Model, Including Examples of Use by STEM Educators in Their Classrooms and PD Facilitators

Application

Application involves defining clear, standards-aligned learning goals and objectives that guide literature-integrated STEM instruction. This step ensures the alignment of STEM concepts and activities with STEM and literacy standards. Teachers select relevant standards, define the STEM concepts embedded in literature, and craft learning goals that connect STEM skills with reading and critical thinking. PD facilitators support educators by modeling how to unpack standards, write interdisciplinary objectives, and use literature to launch STEM-based inquiry. Tools for goal setting and frameworks for integrating standards are provided by PD facilitators.

To set the tone for meaningful integration, the Application phase should start with an introductory activity that demonstrates the role of literature in making STEM content accessible, engaging, and culturally relevant. Literature brings STEM concepts to life by embedding them in narrative contexts that highlight human experiences, real-world challenges, and historical or cultural significance. For example, during the PD kickoff, facilitators might read aloud (via recordings, when appropriate) or share excerpts from biographies, culturally grounded narratives (e.g., *Hidden Figures*), or science-infused novels to spark discussion on how stories can anchor STEM exploration. Participants should then be guided in identifying cross-content standards that align with the text and collaboratively draft interdisciplinary objectives based on the themes presented. In both online and in-person settings, this literature-centered launch can be supported with multimedia and web-based boards for sharing reflections (e.g., Padlet, LMS discussion boards).

During the PD sessions, educators are guided through the process of unpacking standards by identifying verbs and nouns in performance expectations (e.g., "analyze the structure of DNA" or "evaluate a solution") and then mapping them onto literacy outcomes. Facilitators model how to write interdisciplinary objectives that integrate both STEM and ELA standards, such as "Students will evaluate scientific claims in a nonfiction text and construct evidence-based explanations using scientific reasoning." Facilitators also demonstrate how literature, like biographies or science-based narratives, can serve as an anchor for STEM inquiry, such as pairing a novel about space exploration with a physics lesson on motion or using a folktale about engineering to inspire hands-on design challenges. Such activities not only provide a foundation for inquiry-based lesson planning but also highlight literature as a powerful tool for fostering STEM engagement, problem-solving, and inclusivity. When delivering PD online, facilitators can use breakout rooms for collaborative standard deconstruction and shared documents (e.g., Google Docs or Jamboard) for interdisciplinary objective mapping.

Tools such as backward design templates, interdisciplinary objective checklists, and instructional planning frameworks like Understanding by Design (Wiggins & McTighe, 2011) help educators clearly define desired learning outcomes, align assessments with those outcomes, and plan targeted instructional activities that integrate STEM concepts with literary texts. Additional support may include SMART goal graphic organizers or crosswalks between state STEM and ELA standards (National Science Teaching Association [NSTA], 2020). SMART goal graphic organizers can be used to provide a framework for creating clear, measurable, and achievable goals. A crosswalk, in turn, can be used to identify the connection between different sets of standards as many of the skills involved in reading and analyzing text in literature support the development of STEM skills through the stages of critical thinking: (a) comprehension, (b) application, (c) analysis, (d) synthesis, and (e) evaluation. In online learning environments, these tools can be shared via video conferencing or embedded video demonstrations.

Teaching Strategy

The Teaching Strategy phase focuses on selecting literature and instructional strategies to teach the targeted STEM concepts. Teachers analyze the text for embedded STEM ideas, design hands-on STEM tasks, and use instructional practices such as read-aloud, annotation, and modeling to build connections. Emphasizing culturally responsive pedagogy to ensure relevance and accessibility, PD providers facilitate collaborative text analysis and demonstrate strategies that connect story elements to real-world STEM practices. For example, educators are guided through shared readings of high-quality STEM literature, prompting discussion around text structure, vocabulary, and connections to STEM standards. This collaborative unpacking process builds educators' confidence in being able to identify teachable STEM moments within literature. Facilitators also demonstrate strategies that link story elements such as conflict, character motivation, or setting to real-world STEM practices, such as designing an engineering solution or analyzing data trends. For example, a novel about climate change may be used to prompt a graphing activity on temperature patterns or a debate about carbon emissions.

In online formats, PD facilitators can leverage screen sharing with annotation and video conferencing to guide shared readings, prompt discussions, connect shared readings to STEM standards, and demonstrate strategies. Additionally, collaborative platforms like Padlet can be used to host asynchronous discussions. The use of Google Slides with interactive polls or embedded videos can be used to connect text elements with STEM principles.

Table 4.3 provides an example of an activity to help educators explore how literature can serve as a springboard for STEM inquiry and hands-on learning activities.

TABLE 4.3 Example PD Activity for the Teaching Strategy Phase of ATTACK

PD Activity
Purpose: • Educators explain how high-quality literature can serve as a springboard for STEM inquiry and hands-on learning.
Materials to Prepare: • Book copies or excerpts (e.g., *The Most Magnificent Thing, The Boy Who Harnessed the Wind, Ada Twist, Scientist*) • Book study discussion questions • Collaborative shareable thought organizers (e.g., Padlet, Google Doc, Google Slides) • Example STEM-aligned activities

Facilitator Implementation Steps:
• Select 1–2 STEM-aligned texts (e.g., *The Most Magnificent Thing, The Boy Who Harnessed the Wind, Ada Twist, Scientist*).
• Create small groups by grade level or STEM content area.
• Assign guiding questions for discussions.
  • What STEM concepts are evident in the excerpt?
    • "How does literature humanize scientific discovery?"
  • What STEM standard can be addressed by this text? Identify possible learning goals and lesson objectives.
    • "How does the story intersect with STEM concepts and standards?"
  • How does the literary text make the lesson objective relevant to the students you serve (personal, cultural, community, society)?
  • What activities could you implement with your students to help meet the lesson objective?
• Demonstrate and encourage the use of graphic organizers (in a collaborative Padlet or Google Doc) to chart key STEM concepts, activities, standards, and cultural/ethical issues.
  • Encourage groups to capture ideas in the collaborative electronic document (Padlet or Google Doc).
• As needed, facilitate breakout room or roundtable discussions.
• After allowing group time, encourage groups to share their graphic organizers (outline of potential STEM lessons).
• Lead a discussion reflecting on the process.

Culturally Responsive Pedagogy

PD facilitators can emphasize culturally responsive pedagogy by selecting inclusive texts and providing strategies for connecting content to students' cultural contexts and lived experiences (Gay, 2018; Ladson-Billings & Tate, 1995). Culturally sustaining practices such as using stories from diverse voices and facilitating inquiry grounded in local issues ensure that STEM learning is equitable and meaningful for all learners. For example, *Hidden Figures* (Shetterly, 2016) follows three African American women (Katherine Johnson, Dorothy Vaughan, and Mary Jackson) at NASA, who were instrumental in the historical launch of astronaut John Glenn into orbit. Various multimedia can be used. For example, clips from the movie or the text can serve as prompts in discussions of space science such as astronomy, physics, and related concepts (e.g., the formation of galaxies, stars, engineering space equipment). In addition to developing essential STEM skills, these learning experiences support non-majority cultures and women in STEM.

Table 4.4 provides examples of STEM and social-cultural concepts in literature.

Trial/Try-Out

In this phase, PD facilitators use microteaching, peer feedback, and rehearsal scenarios to help educators test and strengthen their delivery. Microteaching is a teacher training

TABLE 4.4 Examples of STEM and Social-Cultural Themes in Literature

Title	*STEM Concept*	*Cultural Relevancy*
The Boy Who Harnessed the Wind (Kamkwamba et al., 2012)	Renewable energy, wind energy principles, engineering mechanics, clean water access	• Highlights resilience and innovation within African communities • Addresses environmental challenges and youth-led problem-solving
Ender's Game (Card & Harris, 1991)	Gravity, weight, mass, gravitational fields, physics of projectile motion	• Explores youth in military STEM training, ethical dilemmas in technology and war, culturally responsive problem-solving strategies
Gattaca (Miller et al., 2010)	Genetics, genetic engineering, space travel, ethical considerations in STEM fields	• Examine societal obsession with genetic perfection, barriers faced by underrepresented groups in STEM, ethical discussions on genetic determinism
The Terminator (Hutson, 1985)	Technology, artificial intelligence, cyborg concepts, bionics	• Explores cultural representations of race, ethnicity, and age in media • Raise ethical questions about technology and humanity
After Earth (David, 2013)	Environmental science, climate change, respiratory health, planetary science	• Connects environmental collapse to current climate issues • Fosters inquiry into air quality and environmental responsibility
Hidden Figures (Shetterly, 2016)	Space science, astronomy, physics, mathematical calculations, engineering	• Highlights contributions of African American women at NASA • Addresses racial and gender equity in STEM • Promotes diverse representation in scientific fields

technique whereby educators practice specific teaching skills for a brief period, often under observation and with feedback. It allows educators to practice short lesson segments in a low-stakes environment. PD facilitators create space for each participant to teach a 5–10-minute excerpt from their literature-integrated STEM lesson while others observe. Structured peer feedback is then provided using a checklist or rubric aligned with the session goals; for example, how effectively the teacher connected the literature to the STEM task. Finally, rehearsal scenarios can simulate classroom conditions, such as managing discussion or differentiating for multilingual learners. This allows educators to refine strategies in real time. These practices build confidence, increase collaboration, and ensure lesson readiness.

In online learning environments, PD facilitators can conduct microteaching through recorded videos where participants share short teaching demonstrations (e.g., GoReact, Flipgrid, Loom). Structured peer feedback can be given via facilitator-created rubrics administered using online surveying or polling tools (e.g., Google Forms, Survey Monkey, or Qualtrics). GoReact, a cloud-based platform, supports skill development via video and time-stamped feedback (SpeakWorks, Inc, 2020). During live sessions, breakout rooms can be created to simulate small-group instructional scenarios.

Table 4.5 shows an example of an activity PD facilitators can use to provide educators with the opportunity to apply insights from the book study to lesson planning and microteaching.

TABLE 4.5 Example PD Activity for the Trial/Try-Out Phase of ATTACK

PD Activity

Purpose:

- Educators apply insights from the book study to lesson planning and microteaching.

Materials to Prepare:

- Planning templates
- Access to book excerpts
- Materials for activity planning
- Feedback forms

Facilitator Implementation Steps:

- Have groups select a book and collaboratively plan a short lesson segment. In online learning environments, group collaboration through breakout rooms.
- Support design of a STEM activity inspired by the text. In online learning environments, provide support by visiting breakout rooms, encouraging participants to use the LMS's chat feature or discussion board, or use a web-based board (e.g., Padlet) to seek support (e.g., pose questions, request support from facilitator).
- Groups deliver a 5–10 min mock lesson demonstrating:
  - Book launch strategy
  - Transition to STEM

Facilitators of online learning can allow participants to present to the whole group via video conferencing or require participants to record and upload their presentation to the LMS's discussion board where their peers and the instructor can comment.

- Use "Glows and Grows" facilitator and peer feedback protocol for refinement. "Glows" represent positive aspects, accomplishments, or things that were done well, and "Grows" indicate opportunities for refinement (Clark, 2017).

Assessment

Assessment refers to the process of measuring learning aligned with the lesson objectives through formative and summative tools. Educators use exit slips, projects, writing prompts, or rubrics to assess comprehension of both STEM content and literary elements. PD facilitators guide educators in creating authentic assessments that measure interdisciplinary learning by providing frameworks for developing performance tasks, student-driven projects, and reflective writing that align with both STEM and literacy standards. This includes modeling how to design rubrics that evaluate not only content mastery but also communication skills such as evidence-based reasoning, clarity of explanation, and integration of literary insights into scientific analysis. For example, a rubric might include criteria for accurate use of scientific vocabulary, thoughtful text-based connections, and collaboration during inquiry tasks. The example lesson planning model in Table 4.2 includes various opportunities for informal and formal assessment activities throughout the lesson (e.g., Turn and Talk, Think-Pair-Share, discussions, debates, quizzes, Quick Write). PD facilitators may also use tools such as performance assessment guides, rubric-building templates, and exemplar student work to help educators align assessments with real-world interdisciplinary expectations.

Facilitators of online learning environments can guide educators in building digital rubrics using technical tools (e.g., Rubistar or Google Forms). Shared documents and folders can facilitate collaborative assessment design. Further, digital exit tickets (e.g., quizzes in the LMS, Google Forms, Nearpod, Quizizz, Kahoot, Poll Everywhere) can be used to model formative assessments and show examples of student work through screen share or shared drives.

Cognitive Reflection

Cognitive reflection involves students and teachers evaluating what they have learned, how they learned it, and how it connects to broader experiences. For example, educators use journals, discussion prompts, or self-assessments to reflect on their learning process and the meaning behind the content.

In this phase, PD facilitators lead educators in reflections on which strategies worked and which ones could be improved (Kaldaras et al., 2024). They accomplish this by embedding reflective practices such as teacher journaling, peer discussion, and student voice protocols directly into PD sessions. For example, facilitators may prompt educators to reflect in writing on how their STEM instruction has evolved through literature or lead structured peer discussions around what surprised them most during implementation. Additionally, PD facilitators encourage the use of metacognitive strategies by modeling how educators can help students set goals, monitor their progress, and self-assess their understanding during a literature-integrated STEM activity. These strategies might include think-aloud, reflection wheels, exit slips, and self-assessment checklists. Emphasizing reflection not only enhances self-awareness but also cultivates habits of mind that support continuous growth in both teachers and students.

Online PD facilitators can (a) embed reflection prompts into a Google Site or LMS discussion forum (e.g., Canvas, Blackboard); (b) use journaling apps or shared documents for teacher reflections; (c) facilitate reflection protocols using activities like "Glow and Grow" (Clark, 2017), "Stars and Wishes" (BetterLesson, n.d.), and "Rose, Thorn, Bud" (University of Colorado Boulder, n.d.) in breakout rooms; and (d) incorporate webtools with interactive boards (e.g., Padlet, Nearpod) to share metacognitive strategies.

Keep/Retain/Generalize

This final phase of the ATTACK model ensures educators retain concepts, use them in new contexts, and apply their learning consistently over time. As such, they plan for ongoing practice, make adaptations for different learners, and integrate literature-STEM lessons into broader units.

PD facilitators support sustainability by helping teachers develop implementation plans tailored to their classroom contexts, including pacing guides, resource inventories, and alignment with curriculum maps. They guide educators through structured planning protocols to define when and how they will use literature-integrated STEM lessons. In addition, they encourage teachers to continuously share strategies with their peers by using collaborative platforms such as Padlet or Google Slides to document adaptations, lesson ideas, and student responses. Finally, facilitators foster long-term growth by connecting educators to follow-up communities of practice either through scheduled virtual meetups, shared discussion boards, or peer observation cycles, where participants reflect on implementation, troubleshoot challenges, and celebrate student learning outcomes.

When presenting online, PD facilitators can provide editable implementation plan templates via shared documents (e.g., Google Docs, Google Slides). They can also (a) utilize interactive walls (e.g., Padlet, Slack) to share an evolving list of strategies; (b) establish a virtual community of practice using groupware (e.g., Google Groups, Microsoft Teams, Canvas discussion boards); or (c) host monthly meetups to support sustained collaboration, troubleshooting, and sharing of success stories via video conferencing (e.g., Zoom, Google Meet). Table 4.6 provides an example of actions and activities when using the ATTACK model to plan a literature-integrated STEM PD.

This final component of the ATTACK model underscores the importance of continued learning and community-building to ensure that instructional strategies become embedded in everyday practice. As illustrated below, this momentum can be further amplified by leveraging online learning platforms to extend learning beyond initial training sessions and foster sustained engagement through virtual collaboration, mentorship, and resource sharing.

Leveraging Online Platforms

Online learning platforms offer a powerful means to deliver and enhance literature-integrated STEM PD by providing flexible, scalable, and inclusive options for reaching educators in diverse geographic and institutional contexts. When structured around the three core practices of (a) collaborative book studies that connect literature to STEM topics, (b) guidance in selecting texts and designing related STEM activities, and (c) opportunities to co-plan interdisciplinary units that integrate literature, STEM inquiry, and hands-on experimentation, online platforms can support meaningful, sustained professional growth. Specifically, these digital environments enable asynchronous and synchronous engagement, offer tools for personalized reflection and feedback, and promote equitable access to high-quality instructional resources. The following are some ways to leverage online delivery, organized by key features of effective PD on literature-integrated STEM.

Collaborative Book Studies That Connect High-Quality Literature to STEM Topics

Using tools like breakout rooms, Google Docs, and discussion boards, educators can engage in guided conversations that explore the STEM themes embedded in narrative texts.

TABLE 4.6 Sample Actions and Activities When Using the ATTACK Model to Plan a Literature-Integrated STEM PD

Phase	*Description*	*Key Components*	*Presenter Actions/Preparation*
A – Application	Establishes purpose and goals of literature-integrated STEM; educators define intended student outcomes.	• Define clear, standards-aligned objectives • Emphasize culturally responsive, interdisciplinary goals • Align objectives with real-world applications and 21st-century skills • Introducing texts that embed STEM concepts	• Prepare sample goals using Bloom's taxonomy and NGSS/CCSS • Provide planning templates for lesson goals • Share a bank of STEM-related texts by grade/ STEM domain • Facilitate discussions on literature's role in STEM (empathy, inquiry)
T – Teaching Strategy	Guides educators through instructional strategies for introducing STEM via literature.	• Criteria for text selection (relevance, diversity, STEM alignment) • Strategies linking narrative themes to STEM inquiry • Culturally responsive practices • Tools for scaffolding and engagement	• Model text analysis to identify STEM themes • Share video/model lessons launching STEM activities • Provide curated literature with aligned STEM prompts • Design collaborative planning sessions (read-aloud → exploration → experiment)
T – Trial/ Try-Out	Teacher's practice applying strategies by developing and presenting portions of lessons with peer feedback.	• Structured mini-teaching and lesson try-outs • Peer review protocols (warm/cool feedback) • Emphasis on low-risk practice and iteration • Reflection on teaching moves and engagement	• Organize breakout groups for co-planning and practice • Distribute feedback forms or checklists • Model constructive feedback processes • Facilitate reflection on successes and support needs
A – Assessment	Explores methods to measure student learning in literature-integrated STEM lessons.	• Rubrics for interdisciplinary skills • Student reflection tools (journals, exit tickets) • Performance-based assessments • Self- and peer-assessment strategies	• Share NGSS/ELA-aligned rubrics • Provide student work samples for analysis • Provide guidance in designing aligned assessments • Offer tools for tracking formative data
C – Closing Reflection and Sustained Support	Debrief and action planning post-cycle to support implementation.	• Reflection on confidence and challenges • Goal setting for classroom application • Ongoing support structures	• Facilitate debrief with reflective prompts • Support goal setting for implementation • Offer virtual coaching, discussion forums, or journal clubs
Optional K – Knowledge Retention and Generalization	Ensures teachers retain and generalize skills across settings and over time.	• Continued application in diverse contexts • Sharing adaptations and practices • Recognition of implementation	• Create shared digital collaboration spaces • Encourage documentation of adaptations • Offer coaching, peer observations, and micro-credentials

For example, platforms such as Google Classroom and Padlet can support ongoing dialogue and idea-sharing across diverse school contexts. Virtual platforms can be used to host journal clubs, webinars, and professional learning communities that connect educators across geographic and institutional boundaries, promoting collaboration and mentorship (ALI Staff, 2023).

Guidance in Selecting Texts and Designing STEM-Integrated Activities

PD providers can use screencasts, instructional videos, and curated resource guides to walk educators through the process of identifying culturally relevant literature and aligning it with hands-on STEM activities. Visual aids, accessibility tools, and digital organizers (e.g., Google Keep, Immersive Reader) can further support multiple means of engagement, representation, and action and expression.

Opportunities to Co-Plan Interdisciplinary Units That Integrate Literature, STEM Inquiry, and Experimentation

Virtual collaboration tools like Jamboard, Google Slides, and shared calendars enable educators to brainstorm, design, and refine cross-curricular units in real time. Further, virtual gallery walks and think-pair-share sessions allow for peer feedback and reflection.

In summary, online learning platforms provide an essential infrastructure for sustaining the impact of literature-integrated STEM instruction beyond the initial PD experience. By supporting collaboration, personalized learning, and equitable access to resources, these platforms help embed effective interdisciplinary practices into educators' everyday instruction. When thoughtfully designed around the three core practices, online PD becomes a powerful tool for building long-term instructional capacity and fostering professional learning communities.

Conclusion

Thoughtfully designed PD creates meaningful opportunities for STEM educators to transform their classrooms into spaces of inquiry, cultural relevance, and interdisciplinary learning by integrating literature into their STEM instruction. The ATTACK model serves as a flexible framework for planning and delivering continuous PD that aligns narrative with scientific discovery, helping educators foster critical thinking, deepen conceptual understanding, and build ethical awareness among students.

By leveraging online platforms, PD providers can scale practices, offering sustained learning experiences that empower educators to collaborate, reflect, and innovate. For example, activities like collaborative book studies, STEM-integrated text selection and activity design, and interdisciplinary unit planning equip teachers to create inclusive, engaging, and standards-aligned STEM instruction.

In closing, literature-integrated STEM instruction is not just about blending disciplines; it is about connecting with students on a human level, sparking curiosity, and inspiring a generation of learners who can think critically, communicate effectively, and solve complex problems with empathy and creativity. As PD providers continue to build capacity in this area, they play a vital role in shaping educators who are prepared to meet the academic and cultural needs of today's diverse student populations.

KEY TAKEAWAYS

- **Integrating Literature with STEM** enhances content accessibility, cultural relevance, and student engagement by embedding real-world narratives into scientific instruction.
- **The ATTACK Model** (Application, Teaching Strategy, Trial/Try-Out, Assessment, Cognitive Reflection, Keep/Retain/Generalize) offers a structured framework for designing both K-12 lessons and PD that effectively blends literature with STEM.
- **Online PD Opportunities** support scalable, flexible, and collaborative professional learning but require intentional strategies to overcome tech barriers, engagement fatigue, and content accessibility issues.
- **Culturally Responsive Teaching** is advanced through literature-integrated STEM by promoting student identity, empathy, inclusion, and critical thinking around ethical and sociopolitical issues in science.
- **Inquiry-Based and Problem-Based Learning** are central pedagogies supported by literature integration, helping teachers and students explore STEM through real-world problems and human-centered stories.
- **The ATTACK Model in PD** guides educators through:
  - Aligning interdisciplinary standards and objectives.
  - Selecting diverse, STEM-rich texts.
  - Designing inquiry-based, hands-on STEM activities.
  - Practicing instructional delivery via microteaching.
  - Creating authentic, formative, and summative assessments.
  - Reflecting on and refining instructional strategies.
  - Retaining and generalizing strategies across different contexts.
- **Literature as a Tool for Equity**: Books like *The Immortal Life of Henrietta Lacks* and *Hidden Figures* provide culturally meaningful contexts to teach about ethics, diversity, and innovation in STEM fields.
- **Persistent Equity Gaps** in STEM education and the STEM workforce – especially for Black, Hispanic, Indigenous, and female students – underscore the need for PD that centers cultural competence.
- **Collaborative Book Studies, Text Selection Guidance, and Interdisciplinary Co-Planning** are the three foundational practices for effective literature-integrated STEM PD.
- **Online Learning Platforms** (e.g., Padlet, Jamboard, Google Docs, Flipgrid) can support sustained collaboration, reflection, resource sharing, and peer feedback in virtual PD environments.

References

ALI Staff. (2023, January 20). *Taking a fresh look at STEM professional development*. Accelerate Learning. https://blog.acceleratelearning.com/stem-professional-development

Alim, H. S., Paris, D., & Wong, C. P. (2020). Culturally sustaining pedagogies: A critical framework for centering communities. In N. S. Nasir, C. D. Lee, R. Pea, & M. McKinney de Royston (Eds.), *Handbook of the cultural foundations of learning* (pp. 261–276). Routledge.

BetterLesson. (n.d.). *Giving and receiving peer feedback*. Retrieved July 31, 2025, from https://teaching.betterlesson.com/strategy/511/giving-and-receiving-peer-feedback

Boon, N. S. (2019). *Exploring STEM competences for the 21st century* (Current and Critical Issues in Curriculum, Learning and Assessment No. 30). UNESCO International Bureau of Education. https://unesdoc.unesco.org/ark:/48223/pf0000368485.locale=enBunch

Bunch-Crump, K., & Hitt, S. B. (2023). Cultural self-awareness: An initial step in home-school partnerships. *Teaching Exceptional Children*, *56*(3), 234–242.

Campbell-Whatley, G. D., Rodriguez, D., & Agrawal, J. (2021). *STEAM meets story: Using adolescent fiction and film to spark deeper learning*. Teachers College Press.

Card, O. S., & Harris, J. (1991). *Ender's game*. Tor.

Chapman, E. (2023, April). STEM readings in English. *Edutopia*. https://www.edutopia.org/article/stem-english-language-arts/

Clark, T. (2017, December 30). *Glows and grows: A positive framing for self-reflection*. https://taviaclark.com/2017/12/30/glow-and-grows/

David, P. (2013). *After earth*. Del Rey.

Delpit, L. (2012). "Will it help the sheep?": Why educate? *About Campus*, *17*(3), 2–9. https://journals.sagepub.com/doi/full/10.1002/aca.21080?casa_token=dY63URza87oAAAAA%3A-qABCSrImVBM2nuTcovc5NrNJL2kOxx8ljsv2ZQZzP-t-bDFyMwnHTr67JxGaQm1jLU09MODt-XKrA

Eden, C. A., Chisom, O. N., & Adeni, I. S. (2024). Cultural competence in education: Strategies for fostering inclusivity and diversity awareness. *International Journal of Applied Research in Social Sciences*, *6*(3), 383–392. https://doi.org/10.51594/ijarss.v6i3.895

English, L. D. (2019). Learning while designing in a fourth-grade integrated STEM problem. *International Journal of Technology and Design Education*, *29*(5), 1011–1032. https://doi.org/10.1007/s10798-018-9482-z

English, L. D., & King, D. (2019). STEM integration in sixth grade: Designing and constructing paper bridges. *International Journal of Science and Mathematics Education*, *17*(5), 863–884. https://doi.org/10.1007/s10763-018-9912-0

Gay, G. (2018). *Culturally responsive teaching: Theory, research, and practice* (3rd ed.). Teachers College Press.

Grangeat, M., Harrison, C., & Dolin, J. (2021). Exploring assessment in STEM inquiry learning classrooms. *International Journal of Science Education*, *43*(3), 345–361. https://doi.org/10.1080/09500693.2021.1903617

Hadaway, N. L., Vardell, S. M., & Young, T. A. (2018). *Science stories: Using children's literature to teach science*. International Reading Association.

Halawa, S., Lin, T. C., & Hsu, Y. S. (2024). Exploring instructional design in K-12 STEM education: A systematic literature review. *International Journal of STEM Education*, *11*, 43. https://doi.org/10.1186/s40594-024-00503-5

Hutson, S. (1985). *The Terminator (film novelization)*. Star Books.

Johnson, P., & Bojorquez, H. (2022, April). Four leverage points for culturally sustaining practices. *IDRA Newsletter*. https://www.idra.org/resource-center/four-leverage-points-or-culturally-sustaining-practices/

Joseph, O. B., & Uzondu, N. C. (2024). Professional development for STEM educators: Enhancing teaching effectiveness through continuous learning. *International Journal of Applied Research in Social Sciences*, *6*(8), 1557–1574. https://doi.org/10.51594/ijarss.v6i8.1370

Kaldaras, L., Wang, K. D., Nardo, J. E., Price, A., Perkins, K., Weiman, C., & Salehi, S. (2024). Employing technology-enhanced feedback and scaffolding to support the development of deep science understanding using computer simulations. *International Journal STEM Education*, *11*, 30.

Kamkwamba, W., Mealer, B., & Zunon, E. (2012). *The boy who harnessed the wind*. Dial Books for Young Readers.

Kim, Y., Kumi-Yeboah, A., & Mohammed, Z. (2024). Promoting culturally responsive teaching in online learning: Experiences and challenges of instructors and students in higher education. *Distance Education*, *45*(3), 385–403. https://doi.org/10.1080/01587919.2024.2345635

Ladson-Billings, G. (2021). Three decades of culturally relevant, responsive, & sustaining pedagogy: What lies ahead? *The Educational Forum*, *85*(4), 351–354. https://doi.org/10.1080/00131725.2021.1957632

Ladson-Billings, G., & Tate, W. (1995). Toward a critical race theory of education. *Teachers College Record*, *97*(1), 47–68.

Lott, K., Wallin, M., Roghaar, D., & Price, Y. (2013). Catch me if you can. Engineering encounters. *Science and Children*, *51*(4), 65–69.

Mahzoon-Hagheghi, M., Yebra, R., Johnson, R. D., & Sohn, L. N. (2018). Fostering a greater understanding of science in the classroom through children's literature. *Texas Journal of Literacy Education*, 6(1), 41–50.

McKinney, S., Tomovic, C., Grant, M., & Hinton, K. (2017). Increasing STEM competence in urban, high poverty elementary school populations. *K-12 STEM Education*, 3(4), 267–281.

Miller, F. P., Vandome, A. F., & McBrewster, J. (2010). *Gattaca*. VDM Publishing.

Milner, H. R. (2012). But what is urban education? *Urban Education*, *47*(3), 556–561. https://doi.org/10.1177/0042085912447516

National Science Teaching Association (NSTA). (2020). *NGSS and common core: Connections and alignments*. https://ngss.nsta.org

Ogodo, J. A. (2022). Developing STEM teachers' cultural competence through an urban teaching curriculum: A cultural border crossing experience. *Journal of Science Teacher Education*, *34*(3), 267–286. https://doi.org/10.1080/1046560X.2022.2046248

Pantaleo, S. (2020). Exploring the intersection of children's literature and science education: Possibilities for teaching and learning. *Literacy*, *54*(3), 145–154. https://doi.org/10.xxxx/lit.2020.xxxx

Pew Research Center. (2021a, April 1). *STEM jobs see uneven progress in increasing gender, racial and ethnic diversity*. https://www.pewresearch.org/science/wp-content/uploads/sites/16/2021/03/PS_2021.04.01_diversity-in-STEM_REPORT.pdf

Pew Research Center. (2021b, December 10). *America's public school teachers are far less racially and ethnically diverse than their students*. https://www.pewresearch.org/short-reads/2021/12/10/americas-public-school-teachers-are-far-less-racially-and-ethnically-diverse-than-their-students/

Rice, D. C. (2012). Using trade books in teaching elementary science: Pedagogical and curricular implications. *Reading & Writing Quarterly*, *28*(3), 253–267. https://www.jstor.org/stable/20205097

Sahito, Z., & Wassan, S. H. (2024). Literature review on STEM education and its awareness among teachers: An exploration of issues and problems with their solutions. *SAGE Open*, *14*(1), 1–21. https://doi.org/10.1177/21582440241236242

Saldívar-Almorejo, M. M., Flores-Herrera, L. A., Rivera-Blas, R., Niño-Suárez, P. A., Rivera-Blas, E. Z., & Rodríguez-Contreras, N. (2024). e-learning challenges in STEM education. *Education Sciences*, *14*(12), 1370. https://doi.org/10.3390/educsci14121370

Santiago, A. (2019). *Focusing on cultural competency in STEM education*. https://informalscience.org/wp-content/uploads/2019/02/Focusing-on-Cultural-Competence-in-STEM-Education.pdf

Shetterly, M. L. (2016). *Hidden figures: The American dream and the untold story of the Black women mathematicians who helped win the space race*. First William Morrow movie tie-in trade paperback edition. William Morrow.

Skloot, R. (2011). *The immortal life of Henrietta Lacks*. Broadway Paperbacks.

SpeakWorks, Inc. (2020). *What is GoReact?* https://help.goreact.com/hc/en-us/articles/360002758171-What-is-GoReact

U.S. Bureau of Labor Statistics. (2024, August 29). Employment projections: 2023–2033 (USD L-24-1776). https://www.bls.gov/news.release/pdf/ecopro.pdf

UNESCO GEM Report. (2024, July 29). *STEM teachers are in short supply*. World Education Blog. https://world-education-blog.org/2024/07/29/stem-teachers-are-in-short-supply/

University of Colorado Boulder. (n.d.). *Rose, Bud, Thorn*. Research & Innovation Office. https://www.colorado.edu/researchinnovation/rose-bud-thorn

Van Hoe, A., Wiebe, J., Rotsaert, T., & Tammy Schellens, T. (2024). The implementation of peer assessment as a scaffold during computer-supported collaborative inquiry learning in secondary STEM education. *International Journal of STEM Education*, *11*, 3. https://doi.org/10.1186/s40594-024-00465-8

Wiggins, G., & McTighe, J. (2011). *The understanding by design guide to creating high-quality units*. ASCD.

Wright, G. A. (2018). Engineering attitudes: An investigation of the effect of literature on student attitudes toward engineering. *International Journal of Technology and Design Education*, *28*, 653–665. https://doi.org/10.1007/s10798-017-9417-0

Zhou, X., Shu, L., Xu, Z., Wang, Z., & Luo, L. (2023). The effect of professional development on in-service STEM teachers' self-efficacy: A meta-analysis of experimental studies. *International Journal of STEM Education*, *10*(1), Article 37. https://doi.org/10.1186/s40594-023-00422-x

5

STEM PROFESSIONAL DEVELOPMENT FOR K-12 TEACHERS

An Online Flipped Learning Model

Bekir Yıldırım

Introduction

Scientific and technological advancements have impacted and continue to impact our lives in many areas, from health to the economy, to education. Specifically, in education, these advancements have facilitated the integration of technology into teaching and learning processes. Such integration not only enhances the quality of teaching and learning but also enriches the overall educational experience, making learning more effective and lasting (Zou et al., 2025).

Various instructional approaches have been integrated into the educational process, including the so-called flipped learning model (Aktı Aslan, 2022), the focus of this chapter. The flipped learning model involves uploading lesson-content videos to an online platform. During the subsequent lesson, practical applications related to the lesson content are then carried out (Bergmann & Sams, 2014). In short,

> FL is a pedagogical approach in which direct instruction is moved from the group learning area to the individual learning area, and the resulting group area is transformed into a dynamic, interactive learning environment where the educator guides students as they apply concepts and engage creatively with the subject matter.
>
> *(The Flipped Learning Network [FLN], 2014, p. 1)*

Watching videos online outside of the classroom allows students to learn at their own pace and review the material whenever they wish, thereby also supporting students with different learning speeds in mastering the lesson content (Abeysekera & Dawson, 2015; Bergmann & Sams, 2012; Enfield, 2013; Thoms, 2012). At the same time, students come to class prepared and ready to actively participate in the lesson, leading to effective learning (Baker, 2000). Further, practical applications carried out during class hours increase the level of interaction by encouraging teacher-student and student-student interaction (Yorgancı, 2020). That is, students make sense of what they have learned after the lesson and apply this knowledge to new situations (Merrill, 2015). Students learn the conceptual learning parts at

DOI: 10.4324/9781003530343-6

their own pace and time followed by practical applications carried out face-to-face during the lesson (Bergmann & Sams, 2012).

> FL can be defined as a teaching strategy or model that combines online, self-paced pre-class learning for basic knowledge acquisition with video lessons and other equivalent solutions, and brings together offline, in-class interactive learning activities facilitated by instructors for knowledge application.
>
> *(Lee, 2023, p. 1182)*

This process ensures integration of theory and practice, which is essential for lasting learning.

Professional Development in STEM Education

Increasingly, teacher professional development is delivered online as this format allows teachers to participate in the training process in a flexible, location- and time-independent manner. Specifically, in online learning platforms, conceptual learning content related to the topic is watched through pre-uploaded videos, followed by practical applications, including classroom practices, also carried out during the online process (Kozikoğlu et al., 2021).

The online flipped learning model uniquely serves these purposes, as conceptual learning is learned asynchronously, whereas the practical part takes place synchronously. That is, the application component occurs online, and teachers interact with each other and receive feedback on the lesson process (Nederveld & Berge, 2015).

With the increasing importance of science, technology, engineering, and mathematics (STEM) education along with a lack of qualified STEM educators, professional development programs are necessary to ensure K-12 teachers have sufficient knowledge and skills to be able to integrate STEM practices into their classrooms (Guzey et al., 2016; Roehrig et al., 2021; Stohlmann et al., 2012). Unfortunately, teachers' work schedules and overall lack of time often hinder them from attending long-term STEM education programs to gain the necessary knowledge (Yıldırım et al., 2022). To overcome this barrier and ensure teachers receive quality training, various educational technological approaches can be utilized to facilitate their professional development (Lu et al., 2022), such as the online flipped learning model.

With the online flipped learning model, teachers can engage in learning activities according to their needs, time, and pace, acquiring conceptual learning knowledge related to STEM education before engaging in actual implementation (Staker & Horn, 2012; Wang & Cheng, 2023). As such, teachers can participate in more interactive activities and receive customized teaching according to their needs (Talley & Scherer, 2013) by gaining greater access to learning materials and learning from different perspectives through diverse resources.

This chapter illustrates how the online flipped learning model can be applied to ensure effective professional development of K-12 teachers in STEM education.

Online Flipped Learning

Despite the obvious benefits of online learning, the question of how to effectively integrate synchronous and asynchronous online learning remains unresolved (Oztok et al., 2013).

Additionally, evidence on how specific learning activities in online environments support or hinder student learning is still insufficient (Nguyen, 2015). A potential response to these challenges may be the online flipped learning model due to its ability to be adapted for fully online settings where face-to-face instruction is not possible.

In the online flipped learning model, unlike the traditional version, face-to-face learning components are conducted online and synchronously (Stöhr et al., 2020). Specifically, the online flipped learning model, first introduced by Bergmann and Sams (2014), consists of three stages: (a) Before the lesson, students watch the lesson content online; (b) During the lesson, students receive feedback and take part in activities; and (c) After the lesson, students assess their learning and apply it to new situations (Merrill, 2015). That is, students acquire conceptual knowledge through online platforms, whereas in-class sessions are dedicated to practical applications (Tucker, 2012). Such a format enables students to come to class better prepared (Honeycutt & Garrett, 2014) and offers an opportunity to compensate for missed lessons related to conceptual understanding (Bergmann & Sams, 2014). Moreover, it supports the development of meaningful learning and higher-order thinking skills (Jia et al., 2022; Jung & Hong, 2020).

The Online Flipped Learning Model in STEM Education

Economic and technological developments and new demands in the business world have led to changes in national policies regarding STEM education and calls for implementing STEM education more widely in K-12 schools (Roehrig et al., 2021), leading to program development efforts, not only in the United States but internationally. For example, in Turkey, the science curriculum was revised in 2018 and 2024 to include engineering-related outcomes (Özyalçın & Avcı, 2024). Similarly, in the United States, a K-12 engineering program has been integrated into the science curriculum (National Academy of Engineering and National Research Council [NAE & NRC], 2014). The underlying goal of these curriculum changes is to provide high-quality STEM education. Moreover, they ensure integration of conceptual learning with engineering practices (Park et al., 2020).

Specifically, the STEM educational approach allows for integrating conceptual learning with practical applications. Additionally, while involving an interdisciplinary perspective, this approach focuses on producing answers to real-world problems (Lin et al., 2021) by generating solutions and creating technological products (Çavdar et al., 2024). STEM education also helps develop skills such as critical thinking, computational thinking, and collaboration (Akcan et al., 2023). Finally, STEM education increases students' interest in future professions and encourages them to pursue these careers (Akcan et al., 2023) by focusing on equipping students with the knowledge and skills required in the business world (Dare et al., 2021).

Online Flipped Learning Model for Teacher Professional Development

Teachers play a crucial role in implementing STEM education practices (Wang & Cheng, 2023), but for K-12 teachers to do so, they need high-quality STEM professional development (Guzey et al., 2016; Roehrig et al., 2021; Stohlmann et al., 2012). Unfortunately, many current professional development programs in this area are inadequate, and the fact that they are typically scheduled on weekdays often prevents teachers from participating (Yıldırım, 2024).

However, educational technologies can be utilized to overcome these barriers (Lu et al., 2022) and, therefore, help K-12 teachers develop professionally in STEM education. Regardless of time and location, the online flipped learning model, in particular, allows K-12 teachers to participate actively, develop more interest in STEM topics, and enhance their professional skills (Fung, 2020), while at the same time learning how to offer more personalized teaching according to their own students' needs (Talley & Scherer, 2013).

In the flipped learning model, an asynchronous part includes the conceptual aspect of STEM education, whereas a synchronous part involves activities and practices related to the application dimensions of the conceptual parts. With a particular focus on how teacher training related to STEM education should be conducted and what should be considered in this process, the STEM Teacher Training Institute model (Yıldırım, 2020) can be delivered in an integrated manner with the flipped learning model, as illustrated in the following.

The STEM Teacher Training Institutes Model

Designed specifically to support professional development in STEM education, the STEM Teacher Training Institute model (Yıldırım, 2020) consists of eight stages: (a) STEM education literacy, (b) identification of needs, (c) setting objectives, (d) content selection, (e) identification of educators and their characteristics, (f) preparing a lesson plan, (g) practice (implementation), and (h) evaluation (Yıldırım, 2020). The model was designed based on the results of a literature review and implementation of a middle school teacher training program (e.g., Du et al., 2019). In particular, the literature review highlighted the importance of identifying needs, selecting appropriate content for teachers, determining qualified educators, and conducting practical applications in the STEM education process (Margot & Kettler, 2019; Nagdi et al., 2018). In the implementation process, teachers from different subject areas were involved, and the model was refined based on their feedback.

The following section presents an example of an online flipped learning model developed for teachers, including program design and application scope based on the STEM Teacher Institute model.

Design and Application of the Online Flipped STEM Learning Model for Teachers

The online flipped STEM teacher training model, which integrates the STEM Teacher Training Institute model (Yıldırım, 2020) and the online flipped learning model, was developed to specifically support the professional development of teachers in STEM education. As such, the model combines the asynchronous and synchronous parts of the flipped learning model with the eight stages of the STEM Teacher Training Institute model (Yıldırım, 2020).

While planning the general framework of the model, a literature review was conducted, focusing on the fundamental knowledge that is necessary for teachers to become STEM literate (Ring et al., 2017; Roehrig et al., 2021; Yıldırım, 2020). After establishing the general framework of the model based on the literature, the needs of the middle school teachers were identified (Dare et al., 2021; Roehrig et al., 2023). Based on this needs assessment, objectives were defined and the content of the training was determined. In line with the identified needs, objectives, and content, an instructor was selected. The instructor was chosen as someone who had experience and expertise in the specific field in which the teachers would be trained.

Accordingly, synchronous and asynchronous course processes were planned to meet the identified needs of the teachers. The final three stages of the STEM Teacher Institute Training model (lesson plan development, implementation, and evaluation) were implemented throughout the synchronous course sessions. Since these stages involve the practical aspects of STEM education, they were carried out during the entire training process.

The following elements should be considered to ensure effective implementation of the model: (a) collaboration, (b) student-centered learning, (c) optimized learning environments, (d) sufficient time for practice, (e) administrative support, (f) information technology support, and (g) reflection and evaluation processes (Bergmann & Sams, 2012; Gülbahar & İlanbey, 2020; McGee & Poojary, 2020; Ünal & Özdinç, 2019). These elements are important for improving the quality of teachers' professional development programs, ensuring their active participation in the process, and enabling them to carry out classroom practices. The absence of these elements may lead to decreased efficiency in applying theoretically taught knowledge. For example, supporting teachers' professional development and the backing of school administration positively affect teachers' professional growth (Gülbahar & İlanbey, 2020). Further, technology-focused learning environments should be collaborative and student-centered (McGee & Poojary, 2020). For instance, in the flipped learning model, the asynchronous part allows learners to progress at their own pace with sufficient time whereas the synchronous part enables teachers to actively participate in the process (Bergmann & Sams, 2012).

Additionally, in terms of the last stage of the model, evaluation, reflective, and evaluative processes are crucial for teachers to analyze the process and determine how well they have achieved their goals. Through reflection and evaluation, teachers can identify their strengths and weaknesses. Therefore, the balanced and holistic inclusion of these elements throughout the training process positively enhances teachers' professional development.

In sum, it is crucial to support the educational process in a collaborative, participant-centered, and needs-oriented manner, while also providing technological support (Ünal & Özdinç, 2019). In technology-supported flipped learning models, such elements are essential for the smooth continuation of the process. Furthermore, collecting feedback on the educational process and conducting evaluations is necessary to ensure the quality of implementation. Additionally, administrative support during the educational process is of critical importance and should be given careful consideration.

The planning of the education process and its connections with the STEM Teacher Institute model are shown in Figure 5.1, which depicts the eight stages that form the structure of the model and indicates when they are applied. The first five stages occur before the lesson, while the remaining three take place during the training process. Figure 5.2 illustrates the integration of the flipped learning model and STEM education before, during, and after the lesson, with the training processes presented cyclically. In short, Figure 5.1 outlines the process of the online flipped STEM teacher training model, while Figure 5.2 demonstrates its application in practice.

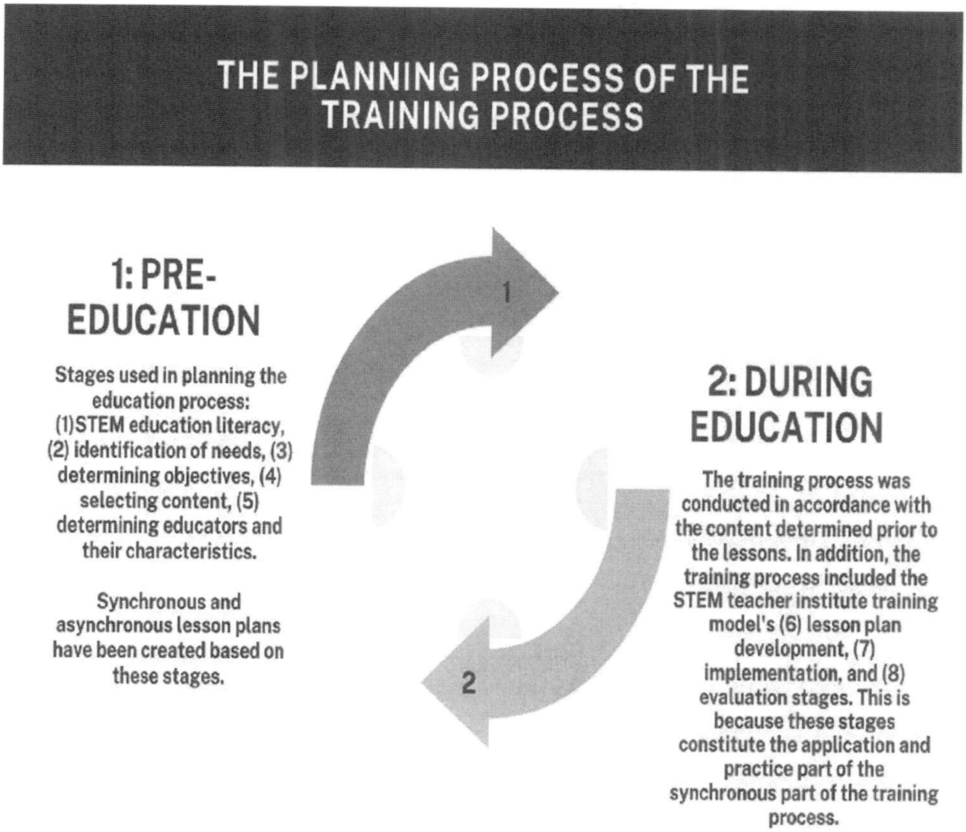

FIGURE 5.1 Planning the educational process and linkages between STEM teacher training institutes

As illustrated in Figure 5.2, STEM education in the online flipped learning model is carried out in three stages as presented in the following.

1 **Pre-Class:** In this stage, videos related to conceptual learning are uploaded to support the professional development of K-12 teachers in STEM education. (To upload pre-lesson videos, a platform forum, such as Google Classroom, is first created.)

   This stage includes the conceptual part of the flipped learning model and forms the foundation for the activities to be conducted during the synchronous lesson. As an example, for a teacher to design a solar oven, they must first possess some fundamental conceptual knowledge. That is, the teacher must understand concepts from science, such as energy, energy transformations, solar energy, light reflection, and conductivity; from mathematics, information such as angles and circles; and from engineering, the processes involved in engineering design.

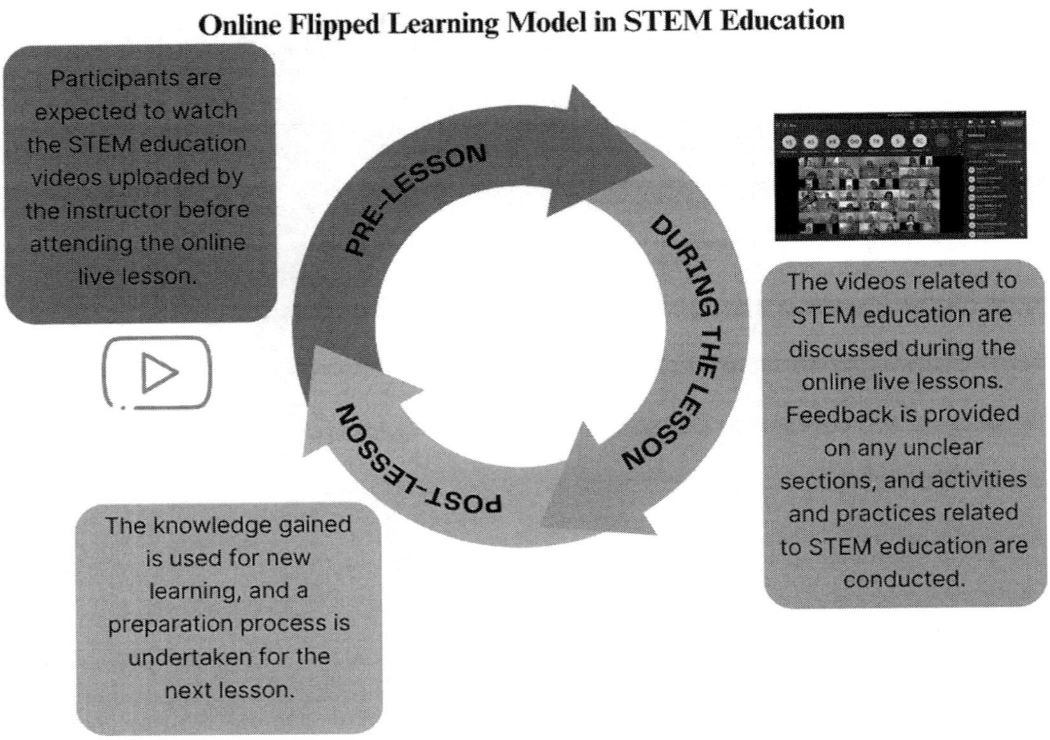

FIGURE 5.2 STEM education in the online flipped learning model

After creating Google Classroom (see Figure 5.3), videos related to STEM education are uploaded to support the asynchronous pre-class part of the online flipped STEM teacher training model (see Figure 5.4; National Aeronautics and Space Administration [NASA]. 2024). The number of videos to be uploaded may vary depending on the scope of the training to be provided. The videos should enable participants to learn conceptual knowledge and be supportive of the activities to be conducted in the synchronous part. For example, as seen in Figure 5.4, 11 lesson recordings related to STEM education

FIGURE 5.3 Class created in Google Classroom

FIGURE 5.4 Conceptual learnings from videos

were uploaded for K-12 teachers, who watched these materials before the class. Once the videos about the basic conceptual knowledge related to the solar oven have been uploaded, an explanatory video on the required materials for the solar oven design should also be provided.

The solar oven activity presented in Figure 5.6 illustrates a STEM activity that teachers can directly implement in their classrooms. It is important that teachers know in advance which scientific concepts are related to the solar oven application and how to apply the engineering design process when conducting the activity during the synchronous session. To that end, teachers are expected to become conceptually well prepared for the solar oven activity in the asynchronous phase. For example, the solar oven designs shown in Figure 5.6 were created by the teachers during the synchronous session using the knowledge they had acquired during the asynchronous phase. Moreover, the teachers placed ice inside the solar ovens they had designed to test them, and then redesigned their models based on the results. The solar oven image presented in Figure 5.6 shows that the teachers had acquired the necessary conceptual knowledge during the asynchronous phase, came to the lesson well prepared, and successfully implemented the activity.

2 **During the Lesson**: During the synchronous lesson (see Figure 5.5), K-12 teachers receive answers to any questions they may have or any points related to the videos that are unclear to them. After the feedback, activities are carried out.

In the synchronous part of the model, no discussions of the conceptual video content watched by the teachers in the asynchronous part take place. Instead, the conceptual knowledge learned is applied through activities and classroom applications. For example, in the case of the solar oven, teachers first acquired the fundamental conceptual knowledge required to build one and then learned about the engineering design processes necessary for creating such a design. In the subsequent synchronous process, a story was presented to the teachers along with a corresponding problem situation to which they were expected to apply the engineering design processes for building a solar oven. The story and problem situation used in the study are provided below.

Merve wanted Ali to design a solar cooker by using solar energy. For this purpose, Merve gave Ali a problem situation and asked him to produce a solution. Ali wrote that he had recently read a news article about how villagers in Zouzugu, Ghana, used the energy

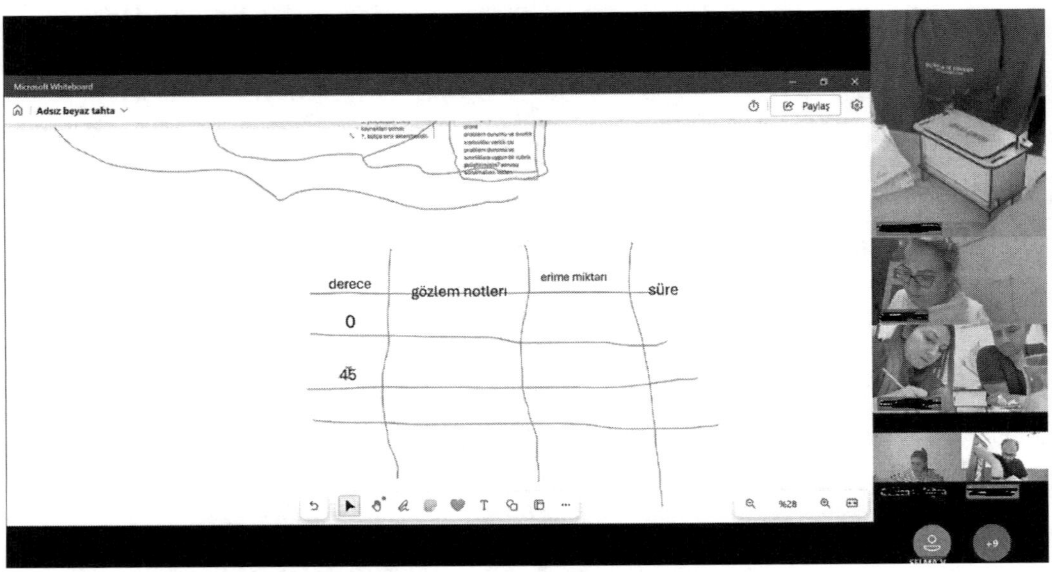

FIGURE 5.5 Synchronous lesson

FIGURE 5.6 Practical applications carried out during the synchronous lessons (Solar oven)

of the sun to pasteurize the water they drink. This news caught his attention, and he started reading it. It stated that the people in Zouzugu did not have sufficient means to pasteurize their drinking water, and because of this, they were getting sick. To solve this problem, the villagers made use of solar energy and designed a solar oven. In this way, they managed to pasteurize their drinking water and started to drink safe water.

Upon reading this, Ali asked, "What is a solar oven?" Merve answered: A solar oven is a device used to cook food or pasteurize liquids by utilizing solar energy. Since the economic situation of the people in Zouzugu is poor, they benefit from solar cookers. Ali thought about what Merve said. When Merve noticed that Ali was thinking, she immediately stepped in. 'Ali, you are also a villager living in another village in Ghana, and the people in your village are also getting sick because they cannot pasteurize the water. So, if you wanted to help the people in your village and design a solar oven to pasteurize water or cook food, what kind of solar oven would you design? The dimensions of the solar oven you design should be $15 \times 15 \times 5$ cm and it should benefit from solar energy to the maximum extent.'

Students, if you were engineers in Ali's place, how would you design a solar oven to pasteurize water or cook food?

Note: Real-life problems are assigned when engineering design processes are applied. A story is provided about the real-life problem to ensure that participants understand the problem situation. The part indicated in black font is the problem situation, while the story describes the problem situation.

The main aim of the synchronous stage is for teachers to apply the knowledge they have learned in the asynchronous stage. In particular, the practical implementation of engineering design processes is important in terms of developing designs such as solar ovens. For example, when the solar oven design created by one of the participants in Figure 5.6 was examined, the teacher tested the design by placing ice inside it. The process proceeds interactively with interaction between teacher and student and between students. For example, Figure 5.5 illustrates the interaction between the instructor and the teachers.

Classroom applications related to STEM education are then carried out. In this example, the activity "We Benefit from Solar Energy: Designing a Solar Oven" was conducted during the live lesson with K-12 teachers (see Figure 5.6).

3 **Post-Lesson:** In this stage, the videos to be watched before the next lesson are discussed, and based on the knowledge gained, teachers are asked to prepare new activity plans. Also, in this stage, the content of the videos required for the next lesson's application is discussed. Furthermore, teachers are asked to plan what kind of classroom activities they can design based on the knowledge they gained from the solar oven activity. While doing this, teachers are specifically asked to write a low-cost, durable, real-life problem-based story that involves the use of solar energy and to reflect on what kind of activities could be designed based on the real-life problem data presented in this story. As a result of this process, the teachers present the activities they have developed and give feedback to each other.

Conclusions and Recommendations

Based on this discussion of the implementation of the reverse learning model in STEM professional development for K-12 teachers, we would like to offer the following recommendations.

The online flipped learning model can be successfully used to expand and enhance professional development programs in STEM education for K-12 teachers (Fung, 2020; Lindeiner-Stráský et al., 2022; Yıldırım, 2024) without the location or time constraints that otherwise often create barriers to professional development.

1 STEM education programs for K-12 teachers should be addressed under two main topics: (a) conceptual learning and (b) practical training (Bergmann & Sams, 2012). These two main components are presented in a holistic way before, during, and after the training process to ensure the highest-quality training. For example, before the training process, flipped learning is carried out through videos uploaded in advance, thus focusing on conceptual learning. During the training (synchronous), activities and practical training related to the conceptual knowledge learned through the previously uploaded videos are carried out. Finally, after the training, evaluation of the knowledge gained in the lesson, as well as planning for the next lesson takes place.

2 A review of the literature revealed that collaboration, student-centered learning, optimized learning environments, time for practice, administrative and technological support, as well as reflection and evaluation play an important role in the effectiveness of the online flipped learning model (Baker, 2000; Yorgancı, 2020). Similarly, they are also important for the effectiveness of STEM education. As a result, when designing an online

flipped learning environment, attention should be paid to the seamless and continuous integration of these elements across the pre-, during-, and post-class phases of this model.

For example, in the current example, in the pre-class phase, conceptual videos are provided within a student-centered learning structure. During this phase, learners can proceed at their own pace and revisit the materials as needed. Since the content is tailored to the specific needs of learners before class, the learning environment is optimized accordingly. During the synchronous phase, participants engage actively in collaborative activities and discussions. This stage allows sufficient time for the practical application of the conceptual knowledge acquired beforehand, contributing to deeper and more meaningful learning experiences. Additionally, both before and after the training, learners receive technology support, which helps to minimize potential technical issues throughout the process. Finally, at the end of the training, participants receive feedback and engage in reflection and evaluation activities.

In closing, as illustrated above, when designing high-quality STEM education using the online flipped learning model, it is essential to adapt the process to the needs and expectations of participating teachers, who, in turn, can more effectively teach their own students.

KEY TAKEAWAYS

- The online flipped learning model supports the professional development of K-12 teachers in STEM education by enabling both asynchronous and synchronous conceptual and practical learning, free from time and space constraints.
- Online flipped learning environments integrate various elements, such as collaboration and technological support.
- To deliver STEM education effectively in online flipped learning environments, the process must be designed according to the needs and expectations of the teachers participating in the training.

References

Abeysekera, L., & Dawson, P. (2015). Motivation and cognitive load in the flipped classroom: Definition, rationale and a call for research. *Higher Education Research & Development*, *34*(1), 1–14. https://doi.org/10.1080/07294360.2014.934336

Akcan, A. T., Yıldırım, B., Karataş, A. R., & Yılmaz, M. (2023). Teachers' views on the effect of STEM education on the labor market. *Frontiers in Psychology*, *14*. https://doi.org/10.3389/fpsyg.2023.1184730

Aktı Aslan, S. (2022). The effect of flipped classroom approach on learning achievement, online self-regulation and interaction in synchronous distance education. *Journal of Educational Technology & Online Learning*, *5*(3), 535–552.

Baker, J. W. (2000). The "classroom flip": Using web course management tools to become the guide by the side. In J. A. Chambers (Ed.), *Proceedings of the 11th International Conference on College Teaching and Learning* (pp. 9–17). Florida Community College at Jacksonville.

Bergmann, J., & Sams, A. (2012). *Flip your classroom: Reach every student in every class every day.* International Society for Technology in Education.

Bergmann, J., & Sams, A. (2014). Flipping for mastery. *Educational Leadership*, *71*(4), 24–29. http://www.ascd.org/el/articles/flipping-for-mastery

Çavdar, O., Yıldırım, B., Kaya, E., & Akkuş, A. (2024). Exploring the nanoworld: Middle school students use TRIZ-STEM in nanotechnology education. *Journal of Chemical Education*, *101*(3), 1049–1061. https://doi.org/10.1021/acs.jchemed.3c01031

Dare, E. A., Keratithamkul, K., Hiwatig, B. M., & Li, F. (2021). Beyond content: The role of STEM disciplines, real-world problems, 21st century skills, and STEM careers within science teachers' conceptions of integrated STEM education. *Education Sciences*, *11*, 737. https://doi.org/10.3390/educsci11110737

Du, W., Liu, D., Johnson, C. C., Sondergeld, T. A., Bolshakova, V. L. J., & Moore, T. J. (2019). The impact of integrated STEM professional development on teacher quality. *School Science and Mathematics*, *119*, 105–114. https://doi.org/10.1111/ssm.12318

Enfield, J. (2013). Looking at the impact of the flipped classroom model of instruction on undergraduate multimedia students at CSUN. *TechTrends*, *57*(6), 14–27.

Flipped Learning Network (FLN). (2014). *The four pillars of F-L-I P™*. http://www.flippedlearning.org/definition. Definition: https://flippedlearning.org/wp-content/uploads/2016/07/FLIP_handout_FNL_Web.pdf

Fung, C. H. (2020). How does flipping classroom foster the STEM education?: A case study of the FPD model. *Technology, Knowledge and Learning*, *25*, 479–507. https://doi.org/10.1007/s10758-020-09443-9

Guzey, S. S., Moore, T. J., & Harwell, M. (2016). Building up STEM: An analysis of teacher-developed engineering design-based stem integration curricular materials. *Journal of Pre-College Engineering Education Research (J-PEER)*, *6*(1), Article 2. https://doi.org/10.7771/2157-9288.1129

Gülbahar, B., & İlanbey, Ö. (2020). Teachers' views on school principal support. *Journal of Social and Humanities Sciences Research*, *7*(62), 3569–3579.

Honeycutt, B., & Garrett, J. (2014). Expanding the definition of a flipped learning environment. *Faculty Focus*. https://www.facultyfocus.com/articles/blended-flipped-learning/expanding-definition-flipped-learning-environment/

Jia, C., Hew, K. F., Bai, S., & Huang, W. (2022). Adaptation of a conventional flipped course to an online flipped format during the COVID-19 pandemic: Student learning performance and engagement. *Journal of Research on Technology in Education*, *54*(2), 281–301. https://doi.org/10.1080/15391523.2020.1847220

Jung, Y. K., & Hong, H. (2020). A theoretical need for applying flipped learning to STEAM education. *Problem-Based Learning*, *7*(1), 42–49. https://doi.org/10.24313/jpbl.2020.00213

Kozikoğlu, İ., Erbenzer, E., & Ateş, G. (2021). The relationship between teachers' flipped learning self-efficacy perceptions and behaviors of supporting learner autonomy. *Journal of Kazım Karabekir Educational Faculty*, *42*, 344–366. https://dergipark.org.tr/tr/pub/ataunikkefd/article/796531

Lee, J. (2023). Flipped learning. In O. Zawacki-Richter & I. Jung (Eds.), *Handbook of open, distance and digital education* (pp. 1179–1196). Springer. https://doi.org/10.1007/978-981-19-2080-6_69

Lin, K. Y., Wu, Y. T., Hsu, Y. T., & Willims, P. J. (2021). Effects of infusing the engineering design process into STEM project-based learning to develop preservice technology teachers' engineering design thinking. *International Journal of STEM Education*, *8*, 1. https://doi.org/10.1186/s40594-020-00258-9

Lindeiner-Stráský, K. V., Stickler, L., & Winchester, S. (2022). Flipping the flipped. The concept of flipped learning in an online teaching environment. *Open Learning: The Journal of Open, Distance and E-Learning*, *37*(3), 288–304. https://doi.org/10.1080/02680513.2020.1769584

Lu, S. Y., Lo, C. C., & Syu, J. Y. (2022). Project-based learning oriented STEAM: The case of micro-bit paper-cutting lamp. *International Journal of Technology and Design Education*, *32*(5), 2553–2575. https://doi.org/10.1007/s10798-021-09714-1

Margot, K. C., & Kettler, T. (2019). Teachers' perception of STEM integration and education: A systematic literature review. *International Journal of STEM Education*, *6*(2), 1–16. https://doi.org/10.1186/s40594-018-0151-2

McGee, E., & Poojary, P. (2020). Exploring blended learning relationships in higher education using a systems-based framework. *Turkish Online Journal of Distance Education*, *21*(4), 1–13. https://doi.org/10.17718/tojde.803343

Merrill, J. (2015). *The flipped classroom: An examination of veteran teacher' practices when flipping their classrooms for the first time* [Unpublished doctoral dissertation]. Texas A&M University.

Nagdi, M. E., Leammukda, F., & Roehrig, G. (2018). Developing identities of STEM teachers at emerging STEM schools. *International Journal of STEM Education*, *5*(36), 2–13. https://doi.org/10.1186/s40594-018-0136-1

National Academy of Engineering and National Research Council [NAE & NRC]. (2014). *STEM integration in K-12 education: Status, prospects, and an agenda for research*. National Academies Press.

National Aeronautics and Space Administration [NASA]. (2024). *Solar oven*. https://science.nasa.gov/learn/heat/resource/solar-oven/

Nederveld, A., & Berge, Z. L. (2015). Flipped learning in the workplace. *Journal of Workplace Learning*, *27*(2), 162–172. https://doi.org/10.1108/JWL-06-2014-0044

Nguyen, T. (2015). The effectiveness of online learning: Beyond no significant difference and future horizons. *MERLOT Journal of Online Learning and Teaching*, *11*, 309–319. https://www.scirp.org/reference/referencespapers?referenceid=2936512

Oztok, M., Zingaro, D., Brett, C., & Hewitt, J. (2013). Exploring asynchronous and synchronous tool use in online courses. *Computers & education*, *60*(1), 87–94. https://doi.org/10.1016/j.compedu.2012.08.007

Özyalçın, B., & Avcı, F. (2024). Examination of engineering design-based science education studies (2012–2024). *Journal of Necmettin Erbakan University Ereğli Faculty of Education*, *6*(2), 992–1009. https://doi.org/10.51119/ereegf.2024.118

Park, W., Wu, J-Y., & Erduran, S. (2020). The nature of STEM disciplines in the science education standards documents from the USA, Korea and Taiwan. *Science & Education*, *29*, 899–927. https://doi.org/10.1007/s11191-020-00139-1

Ring, E. A., Dare, E. A., Crotty, E. A., & Roehrig, G. H. (2017). The evolution of teacher conceptions of STEM education throughout an intensive professional development experience. *Journal of Science Teacher Education*, *28*(5), 444–467. https://doi.org/10.1080/1046560X.2017.1356671

Roehrig, G. H., Dare, E. A., & Ring-Whalen, E. (2021). Understanding coherence and integration in integrated STEM curriculum. *International Journal of STEM Education*, *8*(2), 1–13. https://doi.org/10.1186/s40594-020-00259-8

Roehrig, G. H., Dare, E. A., Wieselmann, J. R., & Ring-Whalen, E. A. (2023). The rise of STEM education: STEM curriculum development and implementation. In R. Tierney, F. Rizvi, & K. Ercikan (Eds.), *International encyclopedia of education* (pp. 153–163). Elsevier. https://doi.org/10.1016/B978-0-12-818630-5.13056-8

Staker, H., & Horn, M. B. (2012). *Classifying K-12 blended learning*. Innosight Institute. http://www.christenseninstitute.org/wp-content/uploads/2013/04/Classifying-K-12-blended-learning.pdf

Stohlmann, M., Tamara, J. M., & Roehrig, G. (2012). Considerations for teaching integrated STEM education. *Journal of Pre-College Engineering Education Research (J-PEER)*, *2*(1), 28–34. https://doi.org/10.5703/1288284314653

Stöhr, C., Demaziere, C., & Adawi, T. (2020). The polarizing effect of the online flipped classroom. *Computers & Education*, *147*, 103789. https://doi.org/10.1016/j.compedu.2019.103789

Talley, C. P., & Scherer, S. (2013). The enhanced flipped classroom: Increasing academic performance with student-recorded lectures and practice testing in a "flipped" STEM course. *The Journal of Negro Education*, *82*(3), 339–347.

Thoms, C. L. V. (2012, November). *Enhancing the blended learning curriculum by using the "flipped classroom" approach to produce a dynamic learning environment*. Paper presented at the 5th International Conference of Education, Research and Innovation (ICERI), Madrid, Spain.

Tucker, B. (2012). The flipped classroom. *Education Next*, *12*(1), 82. https://educationnext.org/the-flipped-classroom/

Ünal, E., & Özdinç, F. (2019). Exploring the experiences of pre-service teachers on technology supported collaborative learning. *MANAS Journal of Social Studies*, *8*(Ek Sayı 1), 794–810.

Wang, T., & Cheng, E. C. K. (2023). Towards effective teacher professional development for stem education in Hong Kong K-12: A case study. In E. C. K. Cheng, T. Wang, T. Schlippe, & G. N. Beligiannis (Eds.), *Artificial intelligence in education technologies: New development and innovative practices*. AIET 2022. Lecture Notes on Data Engineering and Communications Technologies, 154. Springer. https://doi.org/10.1007/978-981-19-8040-4_1

Yıldırım, B. (2024). Flipped TRIZ-STEM: Enhancing teacher training through innovative pedagogy? *Education Information Technologies*, *29*, 10899–10929. https://doi.org/10.1007/s10639-023-12242-1

Yıldırım, B. (2020). A model proposal for teacher training: STEM teacher institutes training model. *PAU Journal of Education*, *50*, 70–98. https://doi.org/10.9779/pauefd.586603

Yıldırım, B., Akcan, A. T., & Öcal, E. (2022). Teachers' perceptions and STEM teaching activities: Online teacher professional development and employment. *Journal of Baltic Science Education*, *21*(1), 84–107. https://doi.org/10.33225/jbse/22.21

Yorgancı, S. (2020). A flipped learning approach to improving students' learning performance in mathematics courses. *Necatibey Faculty of Education Electronic Journal of Science and Mathematics Education*, *14*(1), 348–371. https://doi.org/10.17522/balikesirnef.657197

Zou, Y., Kuek, F., Feng, W., & Cheng, X. (2025). Digital learning in the 21st century: Trends, challenges, and innovations in technology integration. *Frontiers in Education*, *10*, 1–11. https://doi.org/10.3389/feduc.2025.1562391

6

APPLICATION OF THE R2D2 MODEL

Designing Online STEM Teacher Professional Development to Meet Diverse Learning Preferences

Heejung An

Introduction

Online professional development (OPD) has become a popular method for retraining teachers. Even before COVID-19, OPD was gaining momentum due to its accessibility, convenience, flexibility, networking opportunities, and cost effectiveness (Lay et al., 2020). During the pandemic, as many schools closed, numerous additional OPD opportunities emerged to equip educators with the tools and resources necessary to enhance their online teaching practices (Lay et al., 2020). A recent review indicated the effectiveness of OPD in positively affecting student-, teacher-, and classroom-level outcomes, highlighting its enduring importance in teacher education (Morina et al., 2025). Not surprisingly, therefore, OPD has remained a widely adopted and effective delivery format in education.

The shift toward online and blended professional development (PD), whether synchronous, asynchronous, or a combination of both, is driven by factors such as accessibility, affordability, and the ability to tailor learning experiences to individual teacher needs. Despite the positive effects, however, unlike in-person PD, OPD lacks a physical presence, which can lead to participant disengagement (Baran & Correia, 2014; Trust & Horrocks, 2017). Therefore, teacher motivation and engagement are critical factors in ensuring the effectiveness and success of OPD (Borup et al., 2014; Desimone & Garet, 2015). That is, it is essential to move beyond the traditional passive content delivery, such as requiring participants to merely read and respond to discussion board prompts, by incorporating strategies that sustain learner motivation and support long-term retention. Teaching complex or abstract concepts often relies on hands-on experimentations, physical manipulatives, and real-time lab work with observation that must be thoughtfully adapted for a virtual environment (Meylani, 2024). These pedagogical strategies are particularly important in science, technology, engineering, and mathematics (STEM) education, which emphasizes inductive teaching methods such as problem-solving, project- and inquiry-based learning, and guided discovery.

When designing OPD for STEM teachers, various teaching strategies need to be modeled so that the participants can be exposed to STEM-appropriate instructional methods and

DOI: 10.4324/9781003530343-7

best practices that they can replicate in their own classrooms, thereby reinforcing student-centered and interactive approaches to STEM education (Desimone, 2009; Loucks-Horsley et al., 2010). To be able to implement these strategies effectively, teachers must first experience active learning themselves. OPD that integrates multiple modalities not only increases engagement but also enhances motivation and knowledge retention (Bonk & Dennen, 2003), as it allows teachers to interact with content in ways that align with their individual learning preferences.

To support the design of OPD for STEM educators, particularly in addressing individual learning differences, this chapter introduces the Read, Reflect, Display, and Do (R2D2) (Bonk & Zhang, 2008) instructional design model. By fostering dynamic and interactive learning, the R2D2 model ensures that participants not only acquire knowledge but also apply and reflect on it. The model is particularly well suited to STEM instruction because it naturally incorporates the visual representations, experimentation, and application that are fundamental to effective science and mathematics education. In short, it helps OPD instructors improve instructional design, support diverse learners, and develop more effective, engaging STEM courses. To scaffold learning experiences and accommodate varied learning preferences, this chapter integrates evidence-based strategies into the presentation of each of the four R2D2 components.

Strategies for Designing Effective Professional Development

Darling-Hammond et al. (2017) provided a comprehensive synthesis of research on PD found to positively influence instructional practices and student achievement. The authors defined effective PD as structured professional learning that leads to improvements in teacher knowledge, pedagogical practices, and student learning outcomes. Based on an analysis of 35 rigorous studies conducted over the past three decades, their report identified seven key elements commonly found in high-quality PD.

First, effective PD is content-focused, emphasizing subject-specific instructional strategies that are directly tied to classroom practice and aligned with curriculum standards, ensuring relevance and applicability. In online STEM PD, offering discipline-specific modules addresses challenges inherent in teaching complex scientific concepts virtually. Second, active learning is critical; rather than relying on passive lectures, effective PD engages teachers in hands-on activities that mirror the instructional methods they are expected to implement with their own students. This may include virtual simulations, remote laboratory experiences, or teacher-led demonstration using materials that can be found all around us in everyday life. Third, collaboration is essential. Opportunities for teachers to work together foster shared understandings, build professional communities, and support sustained instructional improvement. Online platforms like breakout rooms, collaborative digital white boards, and asynchronous discussion forums can promote collaboration between PD providers and teachers, and between teachers and students. Fourth, modeling of effective practice is critical as it enhances teacher learning through demonstration lessons, video case studies, peer observations, and access to model lesson plans or student work. The fifth element, coaching and expert support, provides individualized guidance from mentors, instructional coaches, or content experts, helping teachers integrate new strategies more effectively. Sixth, feedback and reflection are vital for deep learning; structured opportunities to reflect on instruction, receive feedback, and revise approaches promote critical thinking

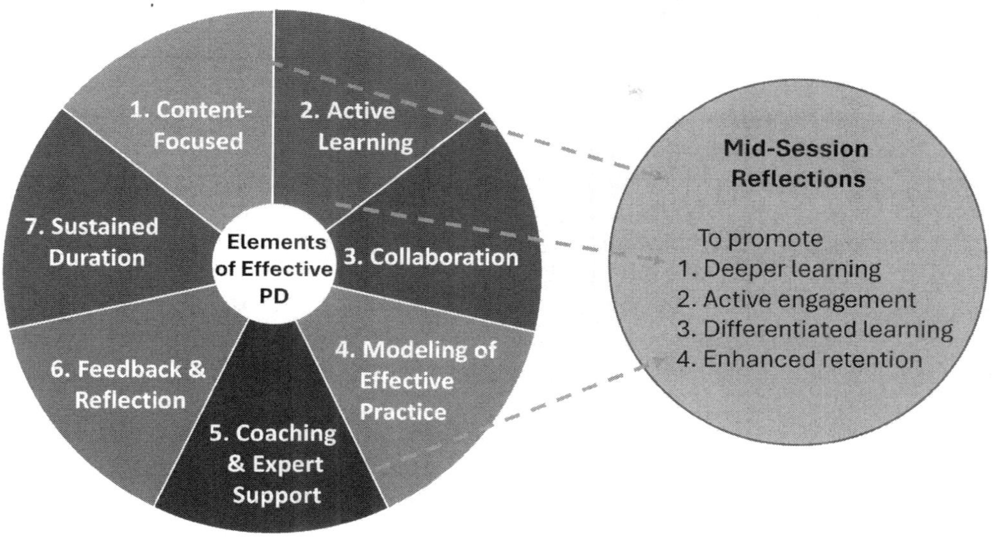

FIGURE 6.1 Key elements and strategies of effective professional development

and continuous improvement. PD facilitators can make this reflection and evaluation process transparent by discussing how they themselves designed and scaffolded the feedback stage. Finally, effective PD is of sustained duration, not a one-time event, but rather a series of sessions spread across weeks, months, or years, allowing time for internalization and refinement of new practices, which in turn supports long-term instructional change and improved student outcomes.

In addition, a growing body of research has highlighted the importance of incorporating mid-session reflections into PD design. First, it promotes deeper learning and internalization, as reflection is a key phase of the learning cycle that enables participants to make sense of experiences and convert them into actionable knowledge (Kolb, 1984). Second, it encourages active engagement, as periodic reflection disrupts passive listening and prompts participants to think critically, discuss ideas, and apply new concepts, an approach that aligns with adult learning theory (Knowles et al., 2015). Third, it supports differentiated and personalized learning by helping participants self-monitor understanding and clarify confusion, thereby making the PD experience more responsive and effective (Guskey, 2002). Finally, reflective pauses enhance retention and transfer, as teachers are more likely to remember and apply what they have learned when they reflect on its relevance and potential application in their own classrooms (Desimone, 2009). Figure 6.1 summarizes the key elements and strategies of effective PD.

Design Principles and Strategies for Online Professional Development in STEM Education

The success of OPD is contingent upon the underlying design principles. Lodhi and Anwar (2025) explored the use of interdisciplinary STEM quests through the Technological Pedagogical Content Knowledge (TPACK) framework (Mishra & Koehler, 2006), and demonstrated how digital learning environments can facilitate subject-matter integration and improve instructional strategies. Further, identifying emerging design principles for online and blended PD in K-12 STEM education, Kowalski et al. (2017) emphasized the importance of interactive content, real-world applications, and opportunities for teacher reflection. Similarly, in an overview of empirical research on OPD, Stavermann (2025) highlighted best practices such as scaffolded learning experiences and the use of digital collaboration tools.

With particular relevance for STEM, Wallace et al. (2021) discussed how OPD supports STEM teachers, particularly during times when traditional resources, such as museums, are inaccessible. The findings of their study suggest that virtual experiences can effectively substitute for in-person learning opportunities, helping teachers develop STEM-related pedagogical skills. Additionally, Miller et al. (2020) examined the design and implementation of an asynchronous OPD course for high school science teachers, focusing on STEM integration and bioinformatics. Findings indicated that well-structured asynchronous PD programs can enhance teachers' ability to incorporate STEM concepts into their instruction. Likewise, Wilson (2023) emphasized the impact of online interactions on teacher learning, particularly in training educators to teach effectively in virtual environments. Specifically, the study points out that meaningful engagement in OPD requires structured interactions that foster collaboration and knowledge sharing among teachers. Similarly, exploring how exposure to PD influences teachers' instructional technology use, Bowman et al. (2020) noted that teachers' beliefs about their technological competencies mediate the effectiveness of such programs.

In summary, the literature on OPD underscores its potential to enhance teacher learning, particularly in STEM education. Key themes emerging from the research include the importance of interactive and collaborative elements, real-world applications, opportunities for teacher reflection, scaffolded learning experiences, and the use of digital collaboration tools. The findings also point to the benefits of both asynchronous and blended learning models, as well as the role of teachers' self-efficacy in using technology.

Instructional Design Framework: R2D2 (Read, Reflect, Display, and Do) Model

The R2D2 model (Bonk & Zhang, 2008) is a flexible framework for online learning that integrates multiple modalities (reading, reflecting, displaying, and doing) to accommodate diverse learning preferences, makes content delivered more enriching for learners, and demonstrates easy-to-apply learning activities for instructors to integrate various technologies for online learning. By allowing engagement without rigidly enforcing a single learning style, the model enhances retention and learner satisfaction. In addition, it aligns with technological advancements, fostering interactive and dynamic learning experiences while providing a structured approach to course design (see Figure 6.2). Bonk and Zhang (2008) explained each quadrant of the R2D2 model as follows.

Reading

The first part of the R2D2 model encourages learners to engage with learning resources by reading materials and listening to online lectures. Since knowledge acquisition typically precedes its application, this stage focuses on building foundational knowledge. In online

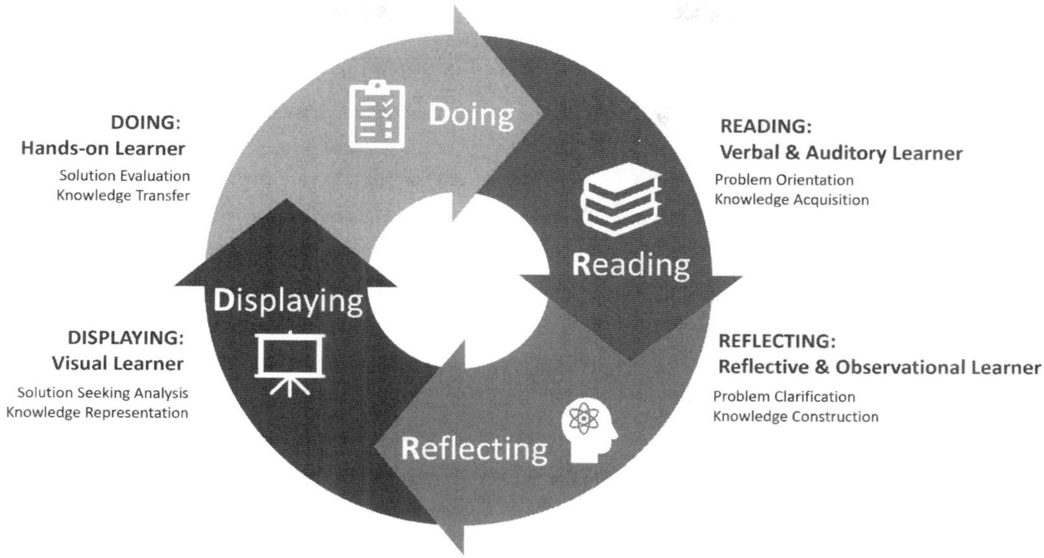

FIGURE 6.2 Read, Reflect, Display, and Do (R2D2) Model

environments, there are various ways in which such knowledge can be acquired, including reading or watching recorded videos.

Reflecting

The second quadrant of the model engages learners in reflecting on what they have learned. The instructor facilitates reflection on the content and encourages self-assessment to help participants check their understanding.

Displaying

The third quadrant encourages learners to represent what they have learned or are in the process of learning through visual representations, depictions, or summaries.

Doing

Finally, the fourth quadrant of the models involves learners applying what they have learned, reflected on, and visualized through practice exercises or real-world applications (Figure 6.2).

Suggested Activities and Instructional Strategies for Each Quadrant of the R2D2 Model

Tables 6.1–6.4 list exemplary learning activities along with applicable online strategies for the Reading, Reflection, Displaying, and Doing quadrants, respectively, of the R2D2 model in the context of science OPD.

TABLE 6.1 Activities, Strategies, and Examples to Use for STEM OPD for the Reading Phase of the R2D2 Model

Learning Activities	*Online Strategies with Examples*	*Takeaways for Teachers as Learners*
Paper-Based Reading (Digital or physical content-specific readings)	• Provide reading materials like PDFs or accessible links. • Use annotation tools such as Perusall (https://www.perusall.com) or Hypothesis (https://web.hypothes.is) to support collaborative reading. • Add journaling or reflection activities. • Organize by theme or module. • Use concept mapping, guided questions, or collaborative synthesis tasks.	Teachers: • Learn how to curate and organize digital reading materials effectively for their students. • Experience how annotation tools can foster student collaboration and deeper engagement with scientific texts. • Understand how to scaffold reading comprehension and gain strategies for making scientific reading more accessible to their learners.
Video-Streamed Lectures (Live, real-time classes delivered over the Internet)	• Use Zoom, Google Meet, or other video conferencing platforms. • Add prompts or reflection activities post-viewing. • Incorporate polls, breakout groups, and shared digital collaboration spaces. • Record sessions for later access (e.g., upload to a YouTube channel or embed in the learning management system (LMS).	• Experience how interactive elements during live-streamed lectures may enhance comprehension and engagement and learn techniques for maintaining student attention and participation during live online sessions.
Recorded Lectures	• Use tools such as Zoom and Camtasia. • Add timestamps, downloadable notes, and comprehension checks using tools like EdPuzzle (https://edpuzzle.com/).	• Learn to design effective comprehension checks and formative assessments within recorded lectures and experience the effectiveness of chunked contents.
Narrated Lectures	• Create audio-narrated slides using tools such as Microsoft PowerPoint, Google Slides with voice-over, Canva Video, or Adobe Express. • Break content into shorter segments for better engagement and understanding.	• Learn the pedagogical benefits of breaking complex STEM concepts into digestible segments with visual and audio.
Guest Expert Talks	• Host live or recorded talks/presentations/interviews. • Provide pre-reading of the guest speaker's work and post-session discussion.	• Gain strategies for connecting classroom learning to real-world STEM applications and careers. • Learn techniques for facilitating meaningful discussion and enhance student motivation by connecting to guest experts' perspectives.

Virtual Tours (Online explorations of real-world or simulated environments)	• Use virtual trip sites:
	  • National Museum of Natural History: https://naturalhistory.si.edu/visit/virtual-tour
	  • Discovery Education: https://www.discoveryeducation.com/community/virtual-field-trips/
	  • NYC Department of Education Virtual Field Trip Collection: https://spep.libguides.com/VirtualFieldTrips
	• Add guiding questions or scavenger hunts.
		• Discover how virtual field experiences provide equitable access to scientific environments and how to design inquiry-based activities that maximize the benefits of virtual exploration.

TABLE 6.2 Activities, Strategies, and Examples to Use for STEM OPD for the Reflecting Phase of the R2D2 Model

Learning Activities	*Online PD Strategies and Examples*	*Takeaways for Teachers as Learners*
Structured Reflection Papers	• Assign reflection papers at the end of the module or session. • Use prompts tied to practice or standards.	Teachers: • Learning how structured reflection deepens student understanding of STEM concepts. • Realize the importance of ongoing reflective writing and collaborative peer feedback during the reflection.
Journaling	• Provide personal or course-based online journaling spaces to support individual reflection and community sharing. • Encourage peer commenting to foster engagement. • Scaffold the journaling process using prompts or exemplars to guide student responses.	

Self-Assessments	• Use quizzes, checklists, self-rating scales, Likert-scale surveys, and rubrics to help learners evaluate their knowledge, skills, or attitudes related to learning goals, utilizing digital tools such as:
	  ◦ Google Forms: https://docs.google.com/forms
	  ◦ Microsoft Forms: https://forms.office.com/
	  ◦ Poll Everywhere: https://www.polleverywhere.com/
	  ◦ Kahoot: https://kahoot.com/
	  ◦ Mentimeter: https://www.mentimeter.com/
	• Include them into reflection journals or learning logs.
	• Encourage goal-setting and progress tracking charts.
	• Integrate pre-, mid-, and post-assessments to monitor growth over time.
		• Develop metacognitive awareness and come to understand how to promote student metacognition and ownership of learning.
		• Learn strategies to guide students' goal-setting and self-monitoring process.

Online Written or Verbal Reflective Discussions	• Conduct threaded or live conversations around readings, video content. • Use learning management systems (LMS)'s discussion forums, Padlet (https://padlet.com/), or Slack (https://slack.com/). • Use protocols like "Think-Pair-Share," sentence stems, or role-based discussion starters.	Develop facilitation skills for guiding meaningful discussion and creating discussion protocols.

TABLE 6.3 Activities, Strategies, and Examples to Use for STEM OPD for the Displaying Phase of the R2D2 Model

Learning Activities	*Online PD Strategies and Examples*	*Takeaways for Teachers as Learners*

Pictures and Diagrams (Static visuals that illustrate key concepts or relationships)	• Create annotated images, science diagrams, or infographics using software programs (examples listed below).
	  • Canva: https://www.canva.com/
	  • Edraw.AI: https://www.edraw.ai/
	  • Chemix: https://chemix.org/
	  • Edraw Max: https://www.edraw-max.com/
	  • Microsoft Office Visio: https://www.microsoft.com/en-us/microsoft-365/visio/flowchart-software
	• Embed into slides or digital journals.
		Teachers:
		• Experience how visual creation and representation tools can transform abstract STEM concepts into concrete, shareable learning artifacts that promote students' understanding.
		• Learn to design authentic assessment opportunities where students communicate scientific thinking through multiple modalities, supporting diverse learners.
		• Develop skills in facilitating student-centered learning where creating visual displays may become a collaborative process that builds scientific communication skills and peer learning.

Drawing and Sketches (Freehand or guided illustrations created to represent ideas)	• Sketch models or diagrams (e.g., life cycles, energy transfer) using software programs (examples listed below) or freehand sketches on paper and embed them into slides or digital journals.
	  • AutoDraw (web-based): https://www.autodraw.com
	  • Kleki (web-based): https://kleki.com
	  • Sketchpad (web-based): https://sketch.io/sketchpad
	  • Microsoft Paint / Paint 3D – Built into Windows
	  • Sketch (web-based): https://sketch.metademolab.com/
	• Use interactive whiteboards during synchronous sessions.	

Animations (Moving visuals that explain processes or concepts over time)	• Show animations related to explanations of processes or concepts using YouTube, pre-existing videos, or by having students create stop-motion animations.	

Concept Maps (Visual representations of relationships between ideas, often hierarchically organized)	• Crate concept maps using tools like:
	  • Cogggle: https://coggle.it/
	  • MindMeister: https://www.mindmeister.com/
	  • Lucidchart: https://www.lucidchart.com/pages
	• Apply as a synthesis task after readings or lectures.	

TABLE 6.4 Activities, Strategies, and Examples to Use for STEM OPD for the Doing Phase of the R2D2 Model

Learning Activities	*Strategies and Examples for OPD*	*Takeaways for Teachers as Learners*

Virtual Simulations or Labs	• Use virtual simulations/labs such as
	  • PhET: https://phet.colorado.edu/
	  • LabXchange: https://www.labx-change.org/
	  • Gizmos: https://gizmos.explore-learning.com/
	• Include guiding or reflection questions and/or data analysis tasks.
		Teachers:
		• Experience how hands-on and virtual experimentation can bridge the gap between theoretical knowledge and practical application.
		• Learn to design authentic learning experiences that engage students as active investigators in real-world problems.
		• Develop strategies for creating instructions and guidelines for boosting students' experiential learning.
		• Understand how collaborative sharing of experimental processes and outcomes builds scientific communication skills and creates learning communities where students learn from each other.

Real Physical Labs	• Assign at-home experiments with safety guidelines. • Use Padlet or discussion boards for sharing procedures and outcomes.	
Project-Based Activities	• Build a model of the scientific phenomenon or system. • Provide structured (with step-by-step instructions) or unstructured (with broader goals or objectives).	

Application of the R2D2 Model and Mid-Session Reflections to Design an OPD Module for Science Teaching

The following lesson is designed to help teachers meet the instructional standards of NGSS: HS-LS1-1, which focuses on DNA structure, base pairing, and how this structure guides protein synthesis; and supports understanding of complementary base pairing that is a key concept in DNA replication. Since this is a professional learning experience for teachers, opportunities for reflection are included at the end of each quadrant. Mid-session reflection at the end of each phase is essential, as it allows for deep learning, active engagement through sharing and meaning dialogue, differentiated learning, and enhanced retention, as depicted in Figure 6.1. For synchronous OPD, reflections may be facilitated during class via video conferencing platforms such as Zoom, using breakout rooms or whole-class discussions. For asynchronous OPD, instructors can create a dedicated discussion forum for each phase.

Figure 6.3 illustrates the OPD module design, which applies the R2D2 model and incorporates a mid-session reflection space called the "Teachers' Lounge." In this space, instructors may post or present prompts for use in breakout rooms or discussion boards. Asynchronous tools such as discussion boards or Padlet (https://padlet.com/) can also facilitate the exchange of ideas, sharing of teaching resources, supporting the collaborative development of a shared repository for future classroom use.

Topic: DNA Structure and Replication.

NGSS: HS-LS1-1: Construct an explanation based on evidence for how the structure of DNA determines the structure of proteins that carry out the essential functions of life through systems of specialized cells.

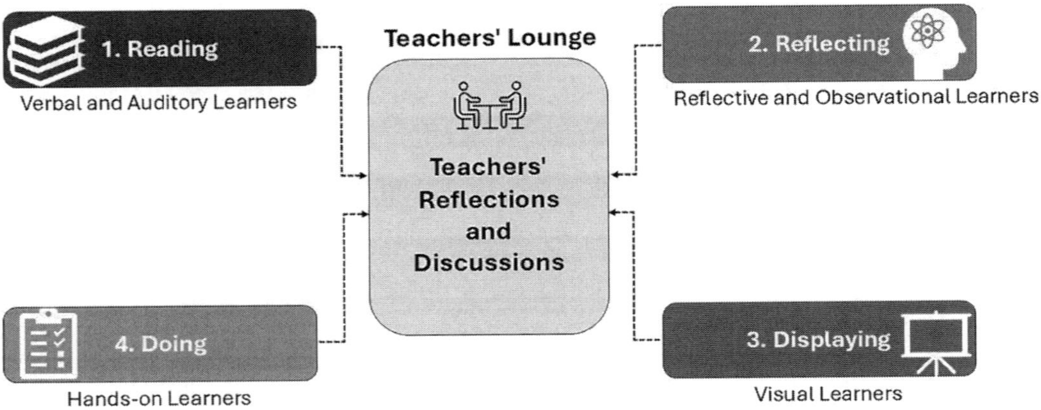

FIGURE 6.3 Online CPD module using the R2D2 model with mid-session reflections (Teachers' lounge)

Reading Phase

At this phase, participating teachers read the provided readings, listen to recorded lectures, and watch videos pertaining to DNA structure and replication. The readings may come from trustworthy websites, textbooks, scholarly or professional articles, video-streamed lectures, recorded or narrated lectures, or videos. (See examples below.) For synchronous CPD, instructors can provide explanations with lecture notes and facilitate discussions related to content. For asynchronous CPD, participants can post questions pertaining to the content.

- Readings:
  - https://flexbooks.ck12.org/cbook/ck-12-middle-school-life-science-2.0/section/3.4/primary/lesson/dna-the-genetic-material-ms-ls/
  - https://www.khanacademy.org/science/ap-biology/gene-expression-and-regulation/replication/a/hs-dna-structure-and-replication-review
  - Video: https://www.youtube.com/watch?v=_POdWsii7AI&ab_channel=AmoebaSisters
- Teachers' Lounge:
  - What are the takeaways pertaining to pedagogical ideas related to the Reading phase to implement and teach your K-12 students?
  - Which parts of the readings do you think your students would find most difficult? Why?
  - Do you have any other readings or videos related to this topic that you might use with your students? Please post them in the Padlet to share.

Reflecting Phase

In the Reflecting phase, participants are asked to reflect on what they have learned through reading or listening activities in the Reading phase. In synchronous CPD sessions, learners can join breakout groups to discuss the topic with a partner and then summarize the discussions back to the main room. It is important to provide prompts related to content comprehension questions to explain the concept. (See examples below.) Participants should respond to the following prompts, either in the discussion board for asynchronous courses or verbally during whole-group or breakout group discussions in synchronous sessions, followed by a class-wide share-out. Participants are also encouraged to assess their knowledge and skills through self-assessment tools such as quizzes, exit tickets, or module check-ins created by the instructor.

- Example prompts for reflection:
  - Describe the structure of DNA.
  - What does nucleotide consist of?
  - What are the base-pairing rules?
  - If one DNA strand reads CCGTAATGCAT, what will be sequence of the complementary strand?
- Example self-assessment quizzes:
  - https://flexbooks.ck12.org/cbook/ck-12-middle-school-life-science-2.0/section/3.4/primary/lesson/dna-the-genetic-material-ms-ls/#review
  - https://glencoe.mheducation.com/sites/0078757134/student_view0/unit3/chapter12/section2/self-check_quizzes.html

Teachers' Lounge

- What are the takeaways pertaining to pedagogical ideas related to the Reflecting phase to implement and teach your K-12 students?
- What misconceptions might students have about the reflection process?
- In what ways can you create opportunities for your students to reflect on what they have learned?
- What types of questions would you ask as a way to self-check what students have learned? Share any resources by posting them to the Padlet.
- What types of questions would you ask to foster deeper learning? Share the questions by posting them to the Padlet.

Displaying Phase

By adopting a lesson published in *The Science Teacher* (Robertson, 2018), participants create a visual representation of DNA replication to help students understand key concepts such as Chargaff's rule (the rule that helped Watson and Crick propose the double-helix model of DNA), which states that the amount of guanine (G) equals the amount of cytosine (C) because G pairs with C, and adenine (A) equals thymine (T) because A pairs with T (Strathern, 1999). The activity also reinforces that DNA replication occurs in the 5′ to 3′ direction, meaning the two strands replicate in opposite directions, ultimately producing two identical copies formed in complementary ways. This approach makes an otherwise abstract concept more accessible and easier to visualize. The sample handout in Figure 6.4 illustrates the directions participants will receive. Instructors ask participants to create a similar diagram on their own sheet of paper. After that, participants are asked to practice

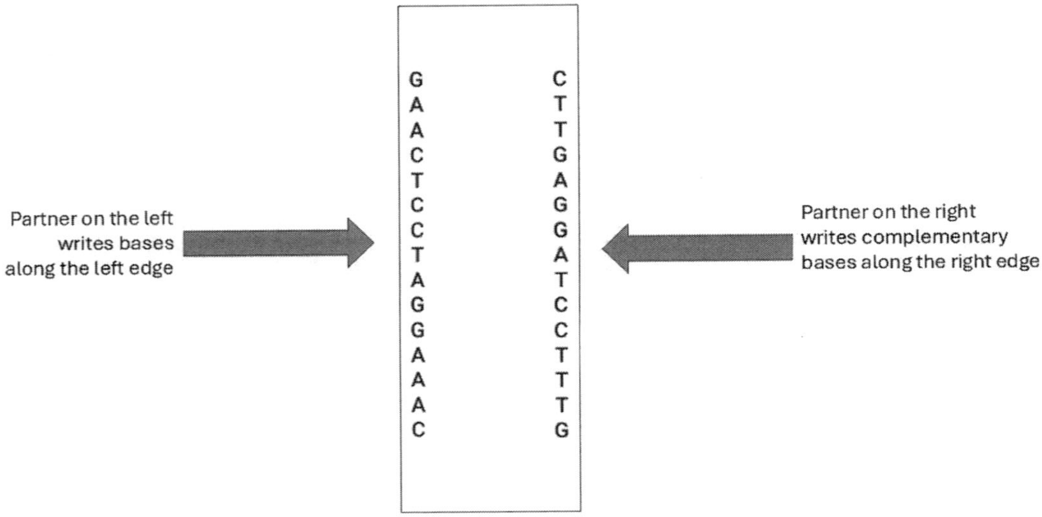

FIGURE 6.4 Handout for an activity in the display stage

applying Chargaff's rule and support this by showing that A always pairs with T, and G always pairs with C in the DNA double helix. See Figure 6.5 for the outcome participants are expected to create. Participants are asked to post their work on the Padlet.

Teachers' Lounge

- What are the takeaways pertaining to pedagogical ideas related to Displaying phase to implement and teach your K-12 students?
- In what other ways can you support your students' understanding of abstract topics through visual representations?

Doing Phase

In the Doing phase, participants apply what they have learned by solving problems and completing various tasks related to DNA construction using a virtual simulation. Additionally, instructors implement project-based learning (PBL), where the project serves as a vehicle for delivering essential knowledge, skills, and concepts aligned with instructional goals (An & Sung, 2024). As part of this approach, participants create a physical 3D model of a DNA helix, enabling them to visualize and apply their learning in a tangible, hands-on manner. In synchronous OPD, teachers can build the model collaboratively by following instructional videos, either as a whole group or in breakout rooms. Alternatively, the instructor may demonstrate the construction process. For asynchronous OPD, the instructor can post a step-by-step video, allowing participants to build the model independently. In either case, instructors should inform participants of the required materials in advance. Once completed, participants are asked to post their models on Padlet (see Figure 6.6)

- Virtual simulation: Building DNA: https://gizmos.explorelearning.com/find-gizmos/lesson-info?resourceId=439

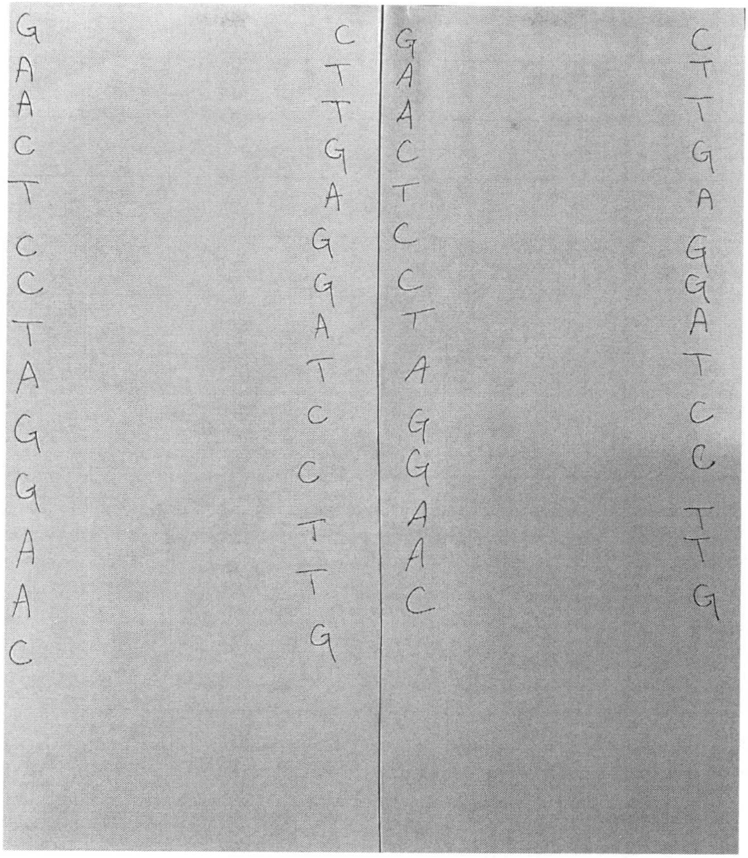

FIGURE 6.5 Outcome of the activity

- Instructional video for creating a 3D DNA helix model:
- https://www.youtube.com/watch?v=BkHKGSDP55Y [supplies needed: four different-colored plastic straws, six skewer sticks, two foam blocks, scissor, ruler, pencil, scotch tape, and black duct tape (not gorilla tape; it is too sticky)].

Teachers' Lounge

- What are the takeaways pertaining to pedagogical ideas related to Doing phase to implement and teach your K-12 students?
- What other types of simulations or hands-on learning activities would you recommend for this phase, where knowledge is constructed through the process of building meaningful tangible artifacts? Share the resources by posting them to the provided Padlet.

At the end of all four phases, a Padlet containing all the resources and ideas shared by participants throughout the R2D2 model is made available to everyone. Participating teachers can then share these collective materials with their own school communities.

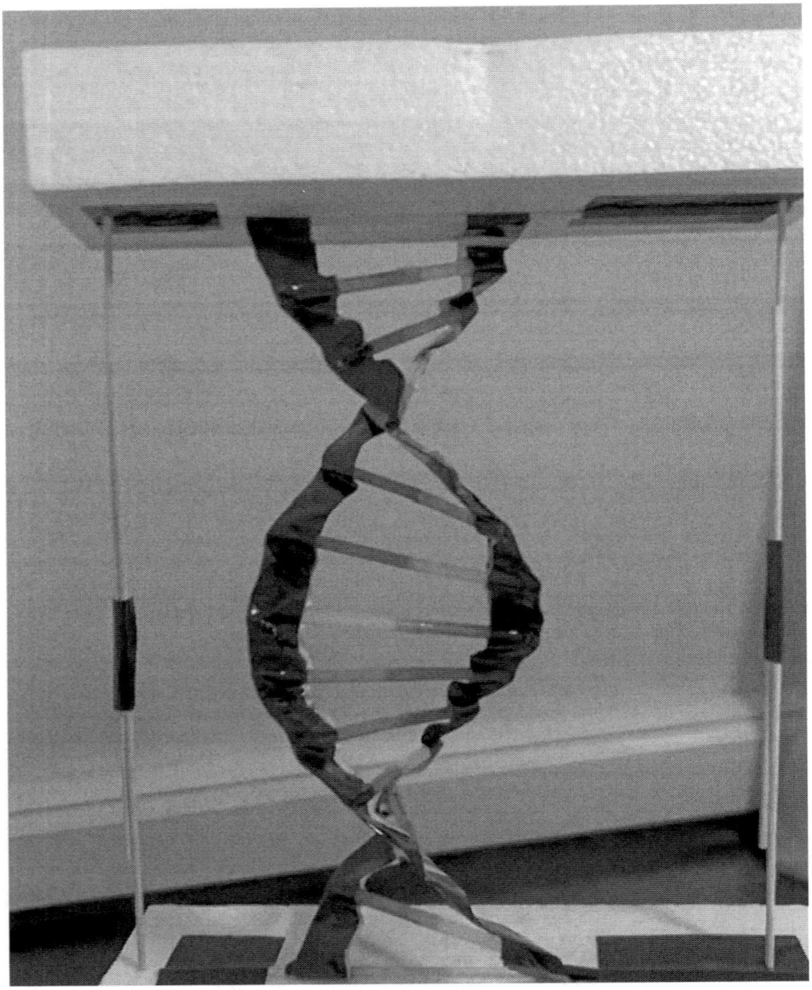

FIGURE 6.6 Completed 3D DNA helix model

Conclusion

Designing effective OPD for STEM teachers requires a deliberate, research-informed approach that enhances instructional practices and promotes student achievement. This chapter reviewed key findings of research on effective PD, including OPD, and introduced the R2D2 model as a flexible, inclusive design framework that accommodates varied learning preferences and promotes active learning in online environments. Embedding evidence-based strategies into each phase of the model enables instructional designers and instructors to create dynamic OPD experiences that reflect best practices in learner-centered STEM education.

The example module on DNA structure and replication presented in this chapter illustrates how the R2D2 model can be applied to support NGSS-aligned instruction while enhancing teachers' pedagogical content knowledge and digital fluency. Ultimately, when teachers are actively and meaningfully engaged in their own learning, they are better prepared to foster the same for their students. As digital learning environments continue to evolve, models like R2D2 offer promising frameworks for designing effective, inclusive professional learning.

KEY TAKEAWAYS

- Effective OPD should be grounded in research-based literature on the effectiveness of PD.
- The R2D2 model is a valuable framework for instructors and instructional designers, as it addresses participants' diverse learning preferences and promotes active learning in an OPD environment.
- To enhance the effectiveness of R2D2-based OPD and better support teachers' classroom needs, it is recommended to incorporate mid-session reflections among participants.

References

An, H., & Sung, W. (2024). The thinkering, making, sharing, and reflecting (TMSR) model: An instructional design model for online collaborative project-based learning. *Educause Review*. https://er.educause.edu/articles/2024/2/thinkering-making-sharing-and-reflecting-a-model-for-online-collaborative-project-based-learning

Baran, E., & Correia, A.-P. (2014). A professional development framework for online teaching. *TechTrends*, *58*(5), 95–101. https://doi.org/10.1007/s11528-014-0791-0

Bonk, C. J., & Dennen, V. (2003). Frameworks for research, design, benchmarks, training, and pedagogy in web-based distance education. In M. G. Moore & B. Anderson (Eds.), *Handbook of distance education* (pp. 331–348). Lawrence Erlbaum Associates.

Bonk, C. J., & Zhang, K. (2008). *Empowering online learning: 100+ activities for reading, reflecting, displaying, and doing*. Jossey-Bass.

Borup, J., West, R. E., Graham, C. R., Archambault, L., & Spring, K. (2014). The adolescent community of engagement framework: A model for research on adolescent online learning. *Journal of Technology and Teacher Education*, *22*(1), 107–130.

Bowman, M. A., Vongkulluksn, V. W., Jiang, Z., & Xie, K. (2020). Teachers' exposure to professional development and the quality of their instructional technology use: The mediating role of teachers' value and ability beliefs. *Journal of Research on Technology in Education*, *54*(2), 123–142.

Darling-Hammond, L., Hyler, M. E., Gardner, M. (2017). *Effective teacher professional development*. Learning Policy Institute. https://doi.org/10.54300/122.311

Desimone, L. M. (2009). Improving impact studies of teachers' professional development: Toward better conceptualizations and measures. *Educational Researcher*, *38*(3), 181–199. https://doi.org/10.3102/0013189X08331140

Desimone, L. M., & Garet, M. S. (2015). Best practices in teachers' professional development in the United States. *Psychology, Society, & Education*, *7*(3), 252–263.

Guskey, T. R. (2002). Professional development and teacher change. *Teachers and Teaching*, *8*(3), 381–391. https://doi.org/10.1080/135406002100000512

Knowles, M. S., Holton, E. F., & Swanson, R. A. (2015). *The adult learner: The definitive classic in adult education and human resource development* (8th ed.). Routledge.

Kolb, D. A. (1984). *Experiential learning: Experience as the source of learning and development*. Prentice Hall.

Kowalski,S. M., Busey,A., Goldsmith,L., & Bates,M., Beilstein,S. O., & Perry,M. (2017). *Emerging design principles for online and blended teacher professional development in K-12 STEM education*. Education Development Center/NSF-supported CADRE report.

Lay, C. D., Allman, B., Cutri, R. M., & Kimmons, R. (2020). Examining a decade of research in online teacher professional development. *Frontiers in Education*, *5*, Article 573129. https://doi.org/10.3389/feduc.2020.573129

Lodhi, S., & Anwar, T. (2025). Exploring interdisciplinary STEMQuest as an opportunity for online teacher professional development using TPACK framework. In R. Jake Cohen (Ed.), *Proceedings of Society for Information Technology & Teacher Education international conference* (pp. 3129–3134). Association for the Advancement of Computing in Education (AACE). https://www.learntechlib.org/primary/p/225919/

Loucks-Horsley, S., Stiles, K. E., Mundry, S., Love, N., & Hewson, P. W. (2010). *Designing professional development for teachers of science and mathematics* (3rd ed.). Corwin Press.

Meylani, R. (2024). Blueprint for the 21st century online learning environment in stem education through a systematic review and qualitative synthesis. *Edelweiss Applied Science and Technology*, *8*(6), 8196–8226. https://doi.org/10.55214/25768484.v8i6.3763

Miller, K. M., Yoon, S. A., & Richman, T. (2020). A social capital design for delivering online asynchronous professional development in a MOOC course for science teachers. *International Journal of Science and Math Education*, *18*(3), 421–440.

Mishra, P., & Koehler, M. (2006). Technological pedagogical content knowledge: A framework for teacher knowledge. *Teachers College Record*, *108*(6), 1017–1054. https://doi.org/10.1111/j.1467-9620.2006.00684.x

Morina, F., Fütterer, T., Hübner, N., Zitzmann, S., & Fischer, C. (2025). Effects of online teacher professional development on teacher-, classroom-, and student-level outcomes: A meta-analysis. *Computers & Education*, *228*, 105247. https://doi.org/10.1016/j.compedu.2025.105247

Robertson, C. (2018). Using modeling to teach DNA replication. *The Science Teacher*, *85*(2), 47–53.

Stavermann, K. (2025). Research in online teacher professional development: A systematic mapping review. *International Journal of Emerging Technologies in Learning (iJET)*, 20(1), 47–67.

Strathern, P. (1999). *Crick, Watson and DNA: The big idea*. Anchor Books.

Trust, T., & Horrocks, B. (2017). "I never feel alone in my classroom": Teacher professional growth within a blended community of practice. *Professional Development in Education*, *43*(4), 645–665. https://doi.org/10.1080/19415257.2016.1233507

Wallace, J., MacPherson, A., Hammerness, K., Chavez-Reilly, M., & Gupta, P. (2021). Pivoting in a pandemic: Supporting STEM teachers' learning through online professional learning during the museum closure. *Journal of STEM Outreach*, *4*(3), 1–13. https://doi.org/10.15695/jstem/v4i3.11

Wilson, M. L. (2023). The impact of technology integration courses on preservice teacher attitudes and beliefs: A meta-analysis (2007–2017). *Journal of Research on Technology in Education*, *55*(1), 25–48.

7

RADIO JOVE CITIZEN SCIENCE TRAINING MODULES

A Self-Paced, Inquiry-Based Environment for Learning Science Through Radio Astronomy

Shing F. Fung, Chuck Higgins, Shirsh Lata Soni, Hilarie Davis, and Carolyn Ng

Introduction: What Is the *Radio JOVE* Project?

Radio JOVE started out as an educational and public outreach project with the goal of inspiring students and the public to learn more about the radio universe through radio astronomy and eventually possibly becoming citizen scientists. As such, it aims to increase science literacy by providing hands-on experience with radio astronomy concepts.

Imagine teachers and students or the general public operating their own radio telescope, making their own observations of the Sun, Jupiter, and the galaxy, sharing their data with practicing scientists at NASA and elsewhere, and actually contributing to the scientific process of acquiring new knowledge. Experiencing the excitement of science discovery first-hand can help learners understand the importance of science, technology, engineering, and mathematics (STEM) skills and help motivate learning. This authentic science experience, made possible in today's classrooms with simple equipment and a great network of *Radio JOVE* citizen scientists, helps put classroom learning in a scientific context to feed curious minds (National Academies of Sciences, Engineering, and Medicine [NASEM], 2018).

Working as partners of the Heliophysics Education Consortium (HEC) and then the NASA Space Science Education Consortium (NSSEC), the two predecessor projects of the NASA Heliophysics Education Activation Team (HEAT; https://science.nasa.gov/learn/heat/), the *Radio JOVE* team quickly realized that *Radio JOVE* was perfectly suited to operate as a citizen science project (as described below) in which participants could help collect and contribute radio observations that could then be analyzed to yield new scientific research results. Major efforts have thus been devoted by the *Radio JOVE* management team since 2015 to develop the standardized *Radio JOVE* 2.1 system described below. One of the design requirements was to keep the system cost low so that it would be affordable to schools or even individuals. A demonstration of the success of *Radio JOVE* as a citizen science project was the discovery of the lightning-generated radio signature called propagation teepee (Fung et al., 2020).

DOI: 10.4324/9781003530343-8

A Unique Entry-Level Practical Radio Astronomy Project

Radio JOVE is an entry-level radio astronomy project. Sponsored originally by the NASA Goddard Space Flight Center, it began operation in the late 1990s as an education and public outreach project with the goal of making radio astronomy, and thus space sciences in general, more accessible to students and the public. Unlike other branches of science, astronomy began as an empirical field of study by humans observing the heavens. That is, since no experiment could be performed directly on the objects being studied, knowledge about celestial objects could only be gained by interpreting observations, incorrectly or correctly (e.g., Ptolemy vs. Copernicus). This remains largely the case to this day except for the parts of the solar system that have been visited by space probes, made possible only since the beginning of space age in the late 1950s. The Voyager spacecrafts, for example, have pushed our limit of *in situ* observations from the solar system into the interstellar medium.

Astronomical observations are by and large made by instruments, such as a signal detector or a camera that captures the signals emanating from distant celestial objects. Due to the great distances of those objects (signal sources) from Earth, a telescope is needed to gather sufficient signal strengths to ensure detection. It is remarkable that everything we are able to learn about celestial objects, and by extension the observable universe, has to come from analyzing the signals we detect remotely from Earth. Knowledge and understanding result only when the astronomical observations can be analyzed objectively and quantitatively using STEM techniques, interpreted and explained in terms of evidence and the physical principles that are established on Earth; that is, the study of astrophysics.

A unique aspect of *Radio JOVE* is that participants can practice radio astronomy without expensive and sophisticated professional equipment as it uses a specially designed and configured radio telescope and receiver system that can be built and operated by amateur participants at modest cost. With such a system, participants can make observations of radio signals broadcasted from the Sun, Jupiter, and the Galaxy. Since the *Radio JOVE* system is optimized to observe or "listen to" radio emissions from Jupiter, the project is named *Radio JOVE* (https://radiojove.gsfc.nasa.gov; Flagg, 2005).

The discovery of radio emissions from Jupiter (Burke & Franklin, 1955; Franklin, 1959, 1983) essentially marked the beginning of the field of planetary radio astronomy, spurring excitement about and interest in studying planetary magnetospheres and culminating in the launch of several space probes (Voyager, Galileo, Ulysses, and Juno) equipped with radio and plasma wave instruments for exploring the magnetospheres of the outer planets.

The Sun is the central body in the solar system. Due to the potential of space weather effects caused by solar activities, such as solar flares and coronal mass ejections (CMEs), (see e.g., https://spaceplace.nasa.gov/spaceweather/en/; https://swe.ssa.esa.int/what-is-space-weather), there are both great scientific interest in and societal concerns about understanding and forecasting how solar activities affect Earth and other bodies in the solar system (Lilensten & Bornarel, 2006). As severe space weather events can pose significant risks to U.S. national security and the economy, the United States Government has issued a National Space Weather Strategy (https://bidenwhitehouse.archives.gov/wp-content/uploads/2023/12/Implementation-Plan-for-National-Space-Weather-Strategy-12212023.pdf) to enable coordinated interagency efforts to forecast and prepare for them. *Radio JOVE* participants can observe remotely the radio signatures of solar activities that are important for understanding and forecasting space weather, including how they can obtain space weather data and contribute to the science of space weather by using STEM knowledge and techniques.

One of *Radio JOVE's* several partners is the Ground Radio Lab (GRL) of the NASA Sun Radio Interferometer Space Experiment (*SunRISE*; https://www.jpl.nasa.gov/missions/sun-radio-interferometer-space-experiment/). SunRISE, a forthcoming NASA space mission, is designed to study solar radio emissions to better understand how the Sun generates intense space weather storms. Its educational outreach component, SunRISE GRL (https://sunrise.umich.edu/), hosted by the Department of Climate and Space Sciences, University of Michigan, USA, is designed to engage students and educators in hands-on radio astronomy activities and real-world scientific investigations. Using *Radio JOVE's* accessible radio telescope kits, the *Radio JOVE*-SunRISE partnership can foster a deeper appreciation of space science by students, teachers, and the general public (Akhavan-Tafti et al., 2024; Soni et al., 2023).

Radio JOVE also has an informal connection with the NASA Juno mission at Jupiter (https://science.nasa.gov/mission/juno/) to monitor Jupiter radio emissions, and advanced user data are being archived at the NASA Planetary Data System archive (https://pds.nasa.gov/). Other important partnerships are the NASA HEAT (https://science.nasa.gov/learn/heat/) where resources and educational materials are shared, and the Society of Amateur Radio Astronomers (SARA; https://radio-astronomy.org/). SARA helps beginners learn about radio astronomy and generously gives grants to cover equipment costs to school groups that qualify. In sum, *Radio JOVE* is part of a large community with many resources to help participants become successful at doing radio astronomy science.

Project Framework: Citizen Science

A NASA Citizen Science Project

Citizen science, or more recently known as *participatory science*, means simply the engagement of public participation in the scientific process in collaboration with domain scientists to address real-world scientific problems (NASEM, 2018). As such, *Radio JOVE* is a NASA-partner citizen science, education, and public outreach project that introduces participants to radio astronomy and provides guidance on scientific practices with STEM applications. Simply put, participants are explorers, and radio astronomy can connect them to the space environment, the Sun, the planets, the galaxy, and the far reaches of the universe. In turn, participants can deduce and learn something new about the universe by applying STEM skills to analyze the observations (https://science.nasa.gov/citizen-science).

Radio JOVE is a community, and as a community it inspires and prepares participants outside the classroom to do NASA science. Through interactions within the community, including NASA scientists and radio astronomy experts, radio enthusiasts are promoted and trained by going through the entire scientific process. As such, the purpose of the *Radio JOVE* training modules is to enable participants to acquire the basic skills and knowledge from building the instrument for making observations, taking and archiving data, and interpreting the observations, to analyzing the data to distill knowledge and understanding. *Radio JOVE* citizen scientists are also encouraged to take part in the final step of the process by reporting their findings in peer-reviewed scientific journals so that their results can be evaluated and validated by the scientific community. Fung et al. (2020) is an example of a peer-reviewed publication of a serendipitous citizen science result from *Radio JOVE*, showing that remote lightning storms may be responsible for the propagation teepee signature often observed by *Radio JOVE* receivers.

Radio JOVE began distributing radio telescope kits in 1999 as a way to teach science through radio astronomy. Using step-by-step instruction manuals, participants built their own single-frequency radio receiver and antenna by soldering parts on a circuit board and assembling the antenna from wires, coaxial cable, and poles, ropes, and stakes to hold the wires in an outdoor field. Data were collected by software running on a computer with optional audio output so participants could hear the radio signals. As quantitative science analysis demands high-quality observations (data), the *Radio JOVE* project developed a calibration system, improved software, and built a data archive (Figure 7.1).

As technology has improved, *Radio JOVE* has become a more science-driven program using commercial radios (software defined radios, SDRs) with multifrequency capability and a frequency-time spectrograph output (see Figure 7.1). These kits do not require soldering, are easier to set up, and have greater capability. More advanced observers have the option of building their own calibration system, thereby improving the scientific value of the data. As a hands-on project, *Radio JOVE* participants build and operate their own radio telescope and conduct radio science studies of the Sun, Jupiter, Earth's ionosphere, and the galactic background. In addition to keeping *Radio JOVE* data at its own data archive (https://radiojove.net/archive.html), plans are to also archive the highest quality *Radio JOVE* data at the NASA Goddard Space Flight Center.

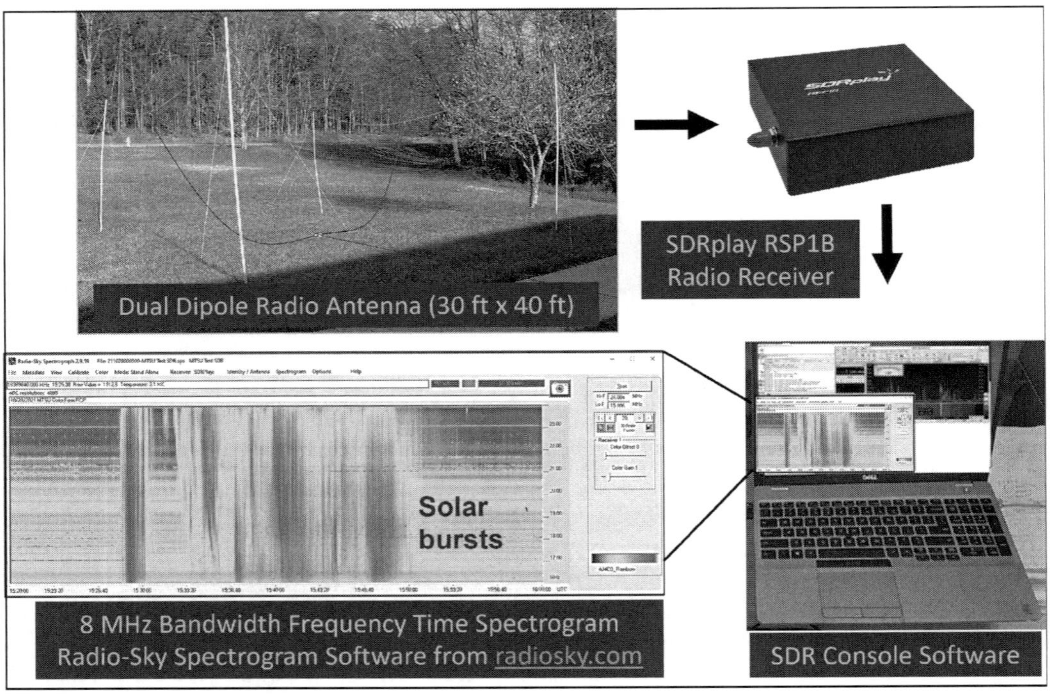

FIGURE 7.1 Typical setup of the Radio JOVE 2.1 telescope

As part of the project, we are interested in science questions like, how do solar flares and CMEs affect interplanetary and magnetospheric plasmas to produce geomagnetic storms (and the auroras) and how do they affect radio waves passing through Earth's ionosphere and disrupt radio communications (a space weather effect). Our radio telescopes operate between 15 and 30 MHz in the high-frequency (HF) radio band using entry-level radio receivers and antennas. Participants use their own radio telescope and engage in a unique mixture of knowledge from the fields of physics, astronomy, engineering, and mathematics. *Radio JOVE* has a worldwide impact with national and international participation of more than 300 registered participants and a friendly and vibrant listserv community. The entry-level equipment is of modest cost, including the radio receiver, antenna, and software, but does not include the cost of a PC-based computer.

Radio JOVE welcomes anyone to join and participate, particularly those who are genuinely interested in making and contributing radio astronomical observations to the performance of science analysis and gaining an understanding of what can be learned from those observations. Radio astronomy is a scientific discipline, so it has its share of technical terms and associated scientific concepts, which means that participants should be open to learning and engaging in discussions using vocabularies that may not be common in their daily lives.

How Radio JOVE *May Be Used by Teachers*

The training modules created by *Radio JOVE*/SunRISE GRL are not formal lesson plans geared toward particular grade levels. Instead, they can be utilized by middle or high school teachers as supplemental materials for STEM education related to space science and radio astronomy. By integrating the modules into club or group activities, teachers can guide students through interactive lessons, videos, and hands-on exercises covering topics such as the fundamentals of radio waves, the Sun's structure and radiation, solar and planetary radio observations, space weather, and the use of radio telescopes.

The modules were developed with multiple purposes in mind: for teachers to (a) learn about *Radio JOVE and SunRISE GRL*, (b) guide their students, (c) use with their students, and (d) collaborate with other teachers. The pilot studies reported below have shown that teachers use the self-paced modules to provide students with a structured yet flexible learning path in space science and radio astronomy. By integrating the modules into their club activities, teachers guided students through interactive lessons, videos, and hands-on exercises covering topics such as the fundamentals of radio waves, solar and planetary observations, and the use of radio telescopes.

What Does One Need to Know Before Starting With the Project?

Basic Science Vocabulary and Concepts

Radio JOVE is an introductory program in radio astronomy and citizen science requiring only high school-level math and science knowledge. Specific knowledge and vocabulary of radio astronomy will be acquired through the project. For example, the first training module only requires an understanding of such prerequisite knowledge as scientific notation, the electromagnetic spectrum (wavelength, frequency), and graphical interpretation of data. Because radio data are recorded by a computer and software, some basic knowledge

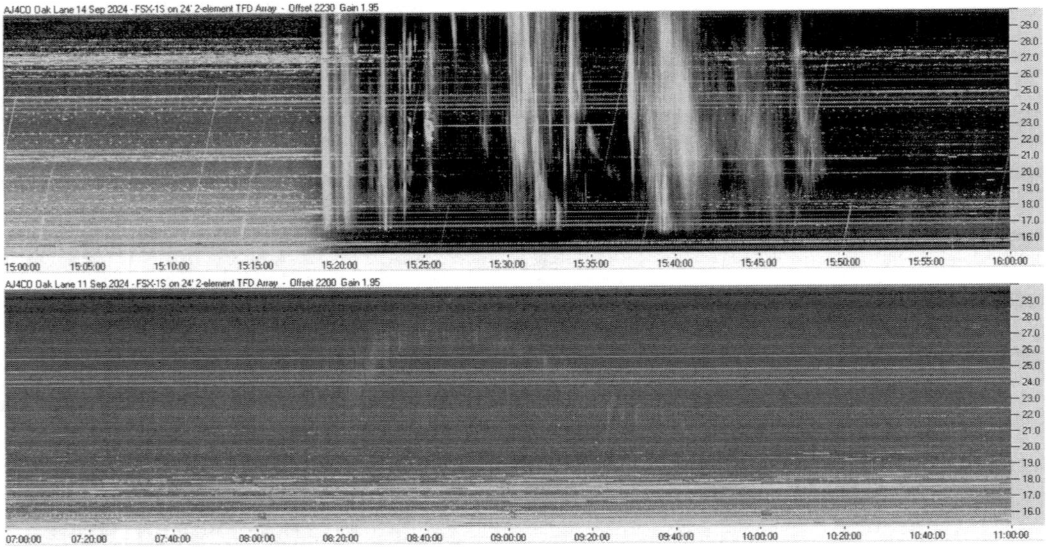

FIGURE 7.2 Two examples of Radio JOVE spectrogram data

of computer and software operation is needed, too (see Figure 7.1). Detailed instructions about software operation are given in step-by-step manuals. Many terms and concepts, especially those that do not typically appear in general science classes, are introduced in the training modules and the corresponding references so participants can learn at their own pace and see how they are used in real applications.

The *Radio JOVE* radio telescope requires setting up an antenna in the field as shown in Figure 7.1 with masts, guy ropes, and stakes and then connecting coaxial cables to the receiver and the computer. No soldering or special knowledge of electrical connections is needed. Many people familiar with amateur HAM radios have more than enough practical knowledge to become excellent radio astronomers. As with most pursuits of knowledge, mainly a passion and understanding on the part of the participants to learn new things are required to achieve success, and *Radio JOVE* provides the foundation and the community to help participants do that. Figure 7.2 shows a couple of examples of solar and Jovian radio burst data that a *Radio JOVE* citizen scientist will likely encounter (Figure 7.2).

Radio JOVE Self-Paced Training Modules

Module Design Considerations

In an increasingly digital world, both teaching and learning can be adapted to diverse learning styles and schedules. Self-paced learning modules offer a flexible, scalable, and inclusive approach to education that accommodates a wide range of learners, from high school students to lifelong learners (Cronin-Golomb & Bauer, 2023; Tullis & Benjamin, 2011). Traditional classroom settings often impose rigid schedules and involve lesson plans that may not align with all participants' availability, readiness, or learning pace. Self-paced modules, on the other hand, empower learners to go through the material at their own convenience without pressure (Torres et al., 2021). This flexibility, in turn, ensures that lessons are accessible to all, including those who may be disadvantaged by conventional educational structures. Self-paced learning promotes long-term engagement by allowing students to revisit content as needed, reinforcing their understanding over time. This continuous interaction with the material leads to better retention and application of knowledge, preparing students for advanced studies or careers in their chosen fields.

Further, with the advancement of digital platforms, self-paced modules can be disseminated to a global audience, breaking geographical barriers. This is particularly crucial for subjects like space science and radio astronomy, for which access to specialized instructional materials may otherwise be limited. By offering these training modules online, we can reach thousands of participants worldwide, fostering a more inclusive and diverse community of learners.

Module Descriptions

The *Radio JOVE* training modules are designed to provide participants with sufficient familiarity with radio astronomy concepts, radio instrumentation, antenna setup, data products, analysis techniques, and their applications in space weather monitoring, so that they can successfully participate in *Radio JOVE*.

The development of the self-paced training modules (Citizen Science Training Modules in both PDF and video formats) began with the design of a framework that applies to each module. After identifying the prerequisite knowledge participants need before proceeding in a given module, the learning objectives for the module are listed. The learning objectives give a clear roadmap for designing the module material, as depicted in Figure 7.3. Participants or module users can quickly review the prerequisite and learning objectives of any module to assess their comfort level with the materials and decide whether they need to

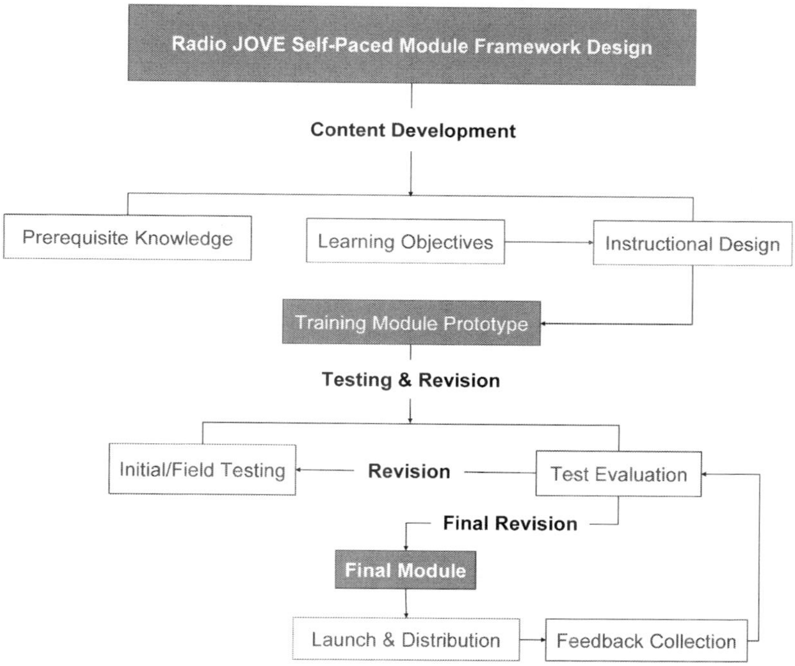

FIGURE 7.3 Radio JOVE training module development workflow

complete the training module or are ready to move on to the next one. (Missing prerequisite knowledge can always be gained later by revisiting earlier modules as needed.) Progressing through the modules in a self-paced manner and with the support of their peers and the *Radio JOVE* community, participants can steadily gain sufficient knowledge to become comfortable with discussing the subject matter with their fellow participants, knowing that everyone has gone through the same path.

The modules are accessible to students and educators worldwide on the *Radio JOVE* website (https://radiojove.gsfc.nasa.gov/citizen_science/training.php), Canvas platform (https://sunrise.umich.edu/), and YouTube (https://www.youtube.com/watch?v=F0tHhrRsRWs). The modules are organized into sections to ensure a logical and consistent flow, starting with an overview of the topic and its specific learning objectives. The modules cover a range of topics, including solar physics, radio astronomy, radio basics, operation instructions, data acquisition, and processing techniques, as well as a brief introduction to citizen science opportunities (Figure 7.4). Practical applications, resources, and case studies are included in each lesson to enhance learning and emphasize real-world relevance.

The modules incorporate engaging multimedia elements such as high-resolution images, interactive visualizations, as well as audio and video clips to help illustrate astronomy, radio phenomena, instrument setups, and data-analysis processes, enhancing learning, making complex concepts more understandable, and bringing the subject matter to life. The modules also include various interactive activities designed to encourage active learning and self-assessment, such as data-analysis exercises, hands-on simulations, and quizzes to reinforce understanding and measure progress. The step-by-step activities and self-assessments guide participants in applying their knowledge to interpret solar radio data and address space science-related challenges.

To ensure usability, the modules use clear and concise language throughout. For example, technical terms and complex concepts are explained in a user-friendly way, making the

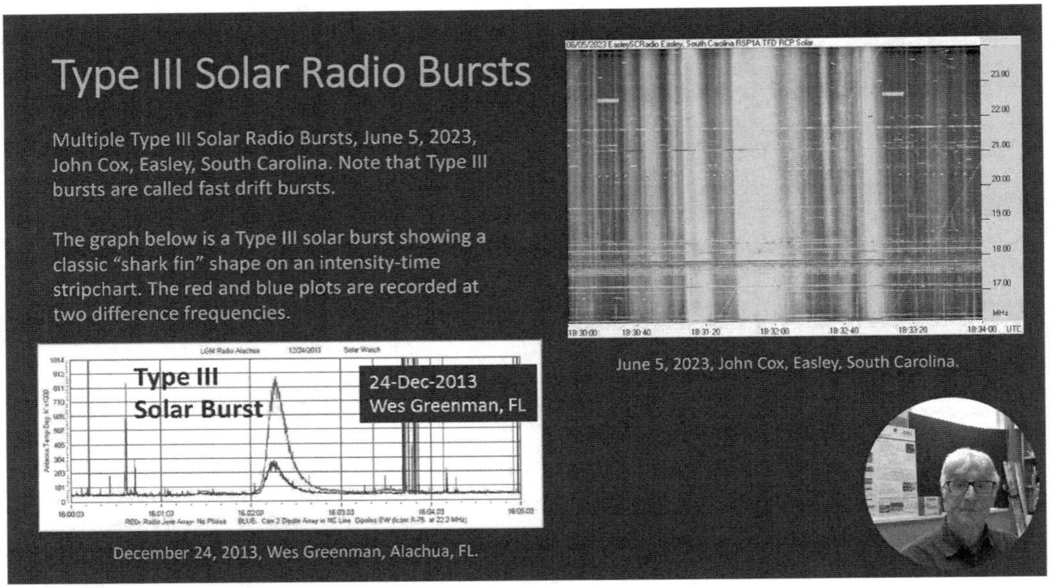

FIGURE 7.4 Page from the video of training module 1.3 on solar radio emissions

material accessible regardless of users' prior knowledge in the field, and definitions, explanations, and examples are provided to enhance comprehension. Further, the materials are presented using text, images, and audio to cater to diverse learning styles and preferences. Additional resources, including research papers, publications, and online databases, are provided to support further learning beyond the core content of the modules; for example, by enabling participants to explore specific topics or pursue advanced studies in solar radio astronomy.

Field Testing Results and Lessons Learned

The development and implementation of the learning modules has been a collaborative and iterative process, heavily informed by field testing with educators and students across different educational levels and geographic locations, as outlined in Figure 7.3. Use of such a diverse testing group ensured that we could evaluate the modules' performance in various learning environments, from traditional classrooms to independent study settings. Participants were encouraged to provide detailed feedback on content clarity, module structure, ease of use, and overall learning experience.

The field testing provided invaluable insights into the effectiveness, usability, and impact of the modules, guiding continuous improvements to meet the diverse needs of our target audience. Thus, the feedback collected from educators and students was instrumental in identifying both the strengths and areas for improvement within the modules, underscoring the importance of adaptability and continuous refinement in educational module development. For example, teachers highlighted the need for additional resources to support differentiated instruction, whereas students appreciated the flexibility offered by the modules but requested more interactive elements to maintain engagement. Both user groups also provided constructive criticism on the pacing and depth of content to help ensure that the modules cater to a wide range of learning speeds and levels of prior knowledge.

Based on the feedback received, several key improvements were made to the modules. For example, we learned that while self-paced learning offers significant benefits, a careful balance of structure and flexibility is required to be effective. Also, the feedback loop between module creators, educators, and students is crucial in ensuring that the content remains relevant, engaging, and accessible to all learners (see Figure 7.3).

Educators using the modules were surveyed about the value of the modules, their use, and effects. The survey was sent to 31 people in July 2024, resulting in 13 responses, which are summarized here. The respondents were from seven institutions (several from the same institution), including four high schools, a university, and an astronomical society. The educators signed up to involve students in radio astronomy, provide alternative data-collection methods during the eclipse, and provide an opportunity to do real science. In total, the educators who responded to the survey had used the *Radio JOVE* modules with 182 students. They rated the six modules from 7.7 to 9.2/10. The majority (12/13) said they could not have done the project without the *Radio JOVE*/SunRISE *GRL* resources.

All the respondents (100%) indicated that they would recommend the project to other teachers. Some illustrative comments were as follows:

- *The SunRISE project is an amazing resource for students to develop hands-on science skills. Students will develop theoretical knowledge in astronomy and learn to apply it through setting up a radio telescope, analyzing data, and drawing conclusions from graphs.*

- *Got to learn more on radio astronomy, the various setups for research, the fascinating results we got. Also got to learn about NASA's Citizen Science Initiative.*
- *An excellent way to introduce yourself to radio astronomy with strong support from the Radio JOVE team to help ensure success.*
- *Our students enjoyed hearing Dr. Higgins share his professional experience, the construction of the radio telescope, and receiving and sharing data from the radio telescope, especially on the solar eclipse days. They were able to contribute to a project much bigger than themselves and our school.*
- *Fascinating results obtained from a self-built radio telescope.*
- *It inspires interest in science.*
- *It has some neat features to pause and teach kids how and why we use this system and what the results mean.*
- *Interesting way to get students into astronomy.*
- *I talked with the school newspaper teacher, and she organized a large article to highlight the project to the whole school (staff and students). Staff have come by to take a look at the antenna.*

Almost 200 students (182) were involved, including 11 in a club setting and 2 in a credit-bearing course. The educators reported great effects on students' interest in science (an average of 9.3 on a 10-point scale), aspirations to a career in science (8.9), engineering experience building the GRL (9.0), confidence in doing science (9.2), and teamwork skills (9.4). Some of the effects reported included:

- *Students were thrilled about having a hands-on activity, building the antenna, and observing. They became more interested in radio astronomy.*
- *It opened their eyes to new science and technology.*
- *They used LOTS of problem-solving skills.*
- *They really got to learn more about radio astronomy, the various setups for research, and the fascinating results we got.*
- *They learned about NASA's Citizen Science Initiative.*
- *It had a profound effect on understanding radio astronomy, Jupiter's interaction with Earth, and working with other interested participants throughout the country.*
- *Fascinating results obtained from self-built radio telescope.*
- *They were able to develop experience in observational astronomy, learn about radio and heliophysics, and acquire hands-on and programming skills.*
- *They improved their soft skills of teamwork, planning, and leadership.*
- *They loved it and were very excited!*

Educators provided additional comments. These included:

- *Great support and kudos to NASA for letting common folks participate in this initiative.*
- *Great organization and support from NASA/Radio JOVE.*
- *Listserv group is great for sharing results and experiences.*

From the survey, we found that the modules were critical to teachers being successful in keeping students engaged with the project (Figure 7.5). For some, the project was an

FIGURE 7.5 Picture showing students from Trinity Christian School, WV, USA, using the Radio JOVE/SunRISE GRL setup

introduction to radio astronomy. For others, it deepened their knowledge. For all, it removed barriers to the actual construction, data processing, and interpretation. To achieve even greater success, respondents recommended providing (a) better tools for semi-automated or automated spectrum analysis to enable easier interpretation; (b) a more automated system for uploading raw data (perhaps along the model of the HamSCI project, https://hamsci.org); (c) more regular teleconferences to improve engagement; (d) clearer scientific project goals; and (e) more feedback on what the data we collected mean.

KEY TAKEAWAYS

- Provide self-paced modules with a structured yet flexible learning path in space science through radio astronomy.
- Educate learners about the electromagnetic spectrum outside the visible range.
- Provide a hands-on experience in building and operating radio telescopes.
- Introduce learners to the possibility of becoming citizen scientists and contributing to science.
- Inspire students to pursue a career in science.

Acknowledgments

This chapter is based on a collaboration between *Radio JOVE* (https://radiojove.gsfc.nasa.gov), SunRISE *Ground Radio Lab* (https://sunrise.umich.edu/), and NASA HEAT (https://science.nasa.gov/learn/heat/). The NASA HEAT is part of NASA's Science Activation portfolio (https://science.nasa.gov/learn/), working to educate people of all ages about heliophysics and inspire future scientists. The authors gratefully acknowledge the suggestions and comments on an earlier draft by Dr. Sten Odenwald.

References

Akhavan-Tafti, M., Soni, S. L., Higgins, C., Fung, S., Lepri, S., Lux, J., Lazio, J., & Romero-Wolf, A. (2024). *Earth and space science.* [Manuscript submitted for publication].

Burke, B. F., & Franklin, K. L. (1955). Observations of a variable radio source associated with the planet Jupiter. *Journal of Geophysical Research*, *60*, 213–217.

Cronin-Golomb, L. M., & Bauer, P J. (2023). Self-motivated and directed learning across the lifespan. *Acta Psychologica*, *232*, 103816. https://doi.org/10.1016/j.actpsy.2022.103816

Flagg, R. S. (2005). *Listening to Jupiter* (2nd ed.). Radio-Sky Publishing Co.

Franklin, K. L. (1959). An account of the discovery of Jupiter as a radio source. *Astronomical Journal*, *64*, 37–39.

Franklin, K. L. (1983, May 4–6). The discovery of Jupiter bursts. In K. I. Kellermann & B. Sheets (Eds.), *Serendipitous discoveries in radio astronomy* (pp. 252–257). In *Proceedings of the NRAO Workshop*, National Radio Astronomy Observatory, Green Bank, WV.

Fung, S. F., Typinski, D., Flagg, R., Ashcraft, T., Greenman, W., Higgins, C., Brown, J., Dodd, L., Reyes, F., Sky, J., Thieman, J., and Garcia, L. (2020). Propagation Teepee: A possible high-frequency (15-30 MHz) remote lightning signature identified by citizen scientists. *Geophysical Research Letters*, *47*, e2020GL087307. https://doi.org/10.1029/2020GL087307

Lilensten, J., & Bornarel, J. (2006). *Space weather, environment and societies.* Springer Dordrecht. https://doi.org/10.1007/1-4020-4332-5

National Academies of Sciences, Engineering, and Medicine (NASEM). (2018). *Learning through citizen science: Enhancing opportunities by design.* The National Academies Press. https://doi.org/10.17226/25183

Soni, S. L., Akhavan-Tafti, M., Higgins, C., & Fung, S. (2023, December 11–15). SunRISE Ground Radio Lab: Engaging 2,500 students and educators in hands-on, heliophysics-focused, STEAM activities. Presentation at the fall American Geophysical Union (AGU) meeting, San Francisco, CA. https://agu.confex.com/agu/fm23/meetingapp.cgi/Paper/1355278

Torres, K. M., Tackett, S., Arrastia-Chisholm, M. C., & Landau, J. (2021). Totally flipping instruction advantages, alternative models, and challenges of a flipped classroom in K-12 and higher education. In I. Jaafar & J. M. Pedersen (Eds.), *Emerging realities and the future of technology in the classroom* (pp. 158–175). IGI Global. https://doi.org/10.4018/978-1-7998-6480-6.ch010

Tullis, J. G., & Benjamin, A. S. (2011). On the effectiveness of self-paced learning. *Journal of Memory and Language*, *64*(2), 109–118. https://doi.org/10.1016/j.jml.2010.11.002

8

TPACK-INFORMED ONLINE CONTINUING PROFESSIONAL DEVELOPMENT

Harnessing Artificial Intelligence Tools for Inquiry-Based STEM Instruction

Heejung An and Sue-Anne Alonso

Introduction

Inquiry-based learning is a core inductive teaching approach, whereby students are presented with an engaging challenge (e.g., a question to be answered, an intriguing picture, an observable event, a situation, or a dataset) that they work through to achieve the intended learning outcome (Prince & Felder, 2007). Pedaste et al. (2015) identified five broad inquiry phases: (a) Orientation; (b) Conceptualization, which includes Questioning and Hypothesis Generation; (c) Investigation, which consists of Exploration, Experimentation, and Data Interpretation; (d) Conclusion; and (e) Discussion, which incorporates Communication and Reflection. Teig et al. (2018) later revised these phases by removing the Orientation phase because it focused more on teacher-initiated approaches for engaging student interest in an investigation than placing the onus on students to take responsibility for their construction of knowledge.

Based on Teig et al.'s (2018) conceptualization, this chapter divides inquiry-based learning into four phases. Figure 8.1 depicts the revised four-phase inquiry model. Notably, the Inquiry cycle is not linear in execution. For example, the Discussion phase (which encompasses Communication and Reflection) can be utilized at any point during the inquiry cycle (known as discussion in-action) or to look back on an inquiry-based learning cycle (known as discussion on-action)

Inquiry-based learning's emphasis on active engagement and problem-solving is closely aligned with the three-dimensional vision of science education outlined in the Next Generation Science Standards (NGSS): Dimension 1 – Science and Engineering Practices (SEPs), Dimension 2 – Disciplinary Core Ideas (DCIs), and Dimension 3 – Crosscutting Concepts (CCCs). That is, the goals of inquiry-based science are "to learn how to do science, learn about the nature of science, and learn science content" (National Research Council [NRC], 2000, p. 1). The NGSS SEPs, in turn, are defined as what scientists do to investigate the natural world and what engineers do to design and build systems (NGSS Lead States, 2013). The practices better explain and extend what is meant by "inquiry" in science and the range of cognitive, social, and physical practices that inquiry requires. Thus, "students engage in

DOI: 10.4324/9781003530343-9

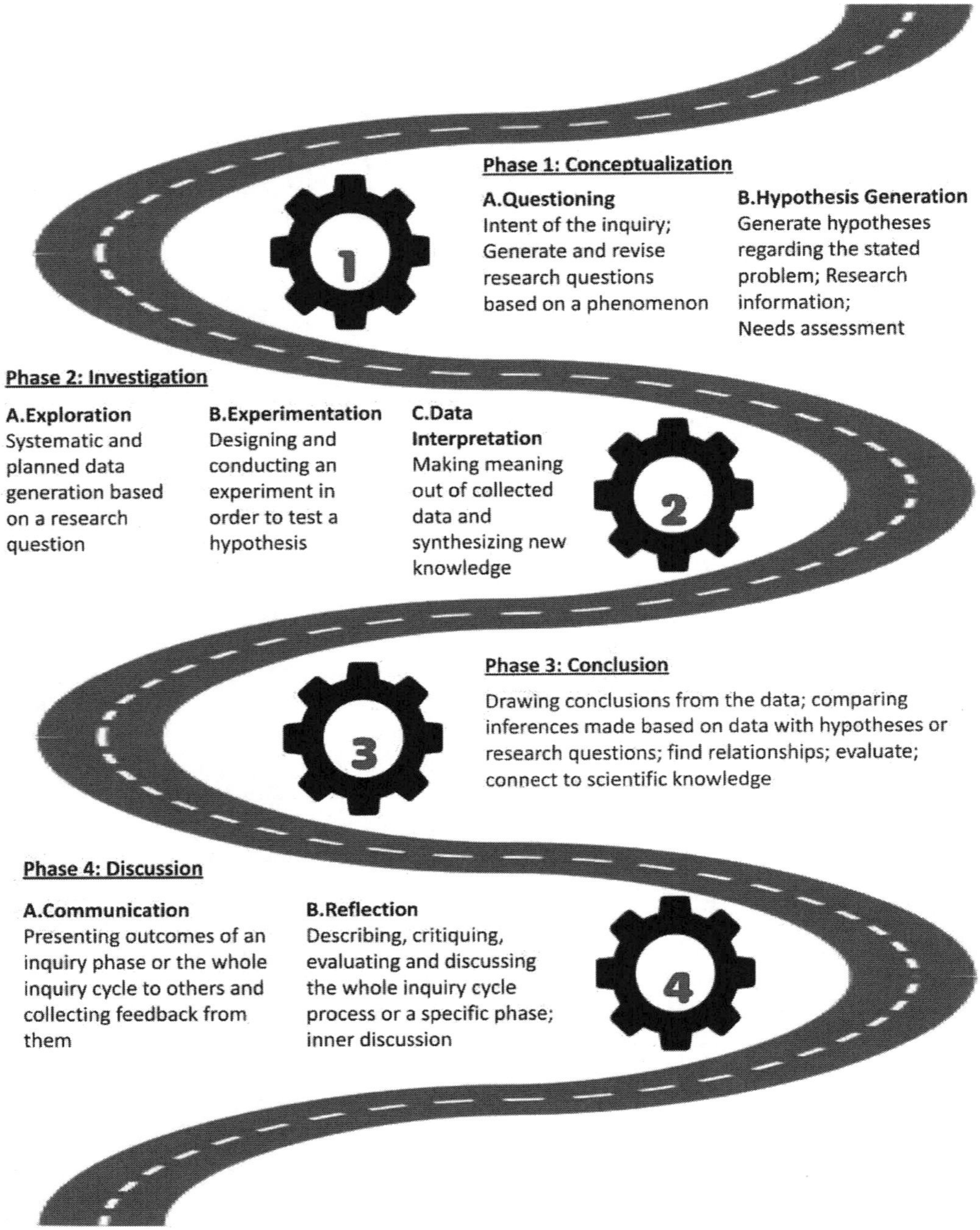

FIGURE 8.1 Revised Four-Phase Inquiry Model

practices to build, deepen, and apply their knowledge of core ideas and crosscutting concepts" (NGSS, n.d., para 3). The term *practices*, rather than *skills*, was deliberately chosen to emphasize that scientific inquiry requires both conceptual understanding and applied competence (NRC, 2012).

More recently, new opportunities in education have been provided by the rise of the Fourth Industrial Revolution (IR4), which is seen as following the Third Industrial Revolution – the Information Age. Reflective of the rapid technological advancements taking place in the 21st century, IR4 consists of an amalgamation of physical, digital, and biological tools, including big data, robotics, the Internet of All Things, and AI (Arucu, 2021; Nel & Kayembe, 2019). In particular, artificial intelligence (AI) tools are of interest to STEM education researchers and practitioners as a means of developing contemporary, engaging educational environments beyond traditional teacher-centric pedagogy to provide "learning experiences that are embodied, project-based, inquiry-driven, collaborative, and open-ended" (Niemi et al., 2023, p. 337).

Despite these potential advantages, a systematic review of professional development opportunities related to AI use in education found a mismatch between training offered in this area and the needs of teachers (Chiu et al., 2023). That is, many teacher preparation programs lack AI training (Delello et al., 2025; Park et al., 2023; Weiner et al, 2024), while others focus on plagiarism issues as opposed to the effective use of AI in classrooms (Weiner et al., 2024). Additionally, many traditional professional development sessions survey a plethora of platforms, without tailoring instruction to the particular needs of teachers in specific subject areas. Against this background, Du et al. (2024) noted that AI training for K-12 teachers is still in its infancy and echoed a critical need for enhancing teachers' professional learning in this area.

To fill this gap, this chapter will exemplify an online continuing professional development (CPD) on how AI tools can be meaningfully leveraged to support inquiry-based science teaching. We will specifically address how AI tools can be incorporated to support three SEPs – Practice 1: Asking Questions; Practice 2: Developing and Using Models; and Practice 4: Analyzing and Interpreting Data.

Practice 1: Asking Questions

The first SEP to be addressed is the cornerstone of inquiry-based science learning, and coincides with Phase 1 of the inquiry process: Conceptualization. The benefits of this practice include "enhancing students' comprehension, academic achievement, motivation, question-generation abilities, problem-solving abilities, cognitive and metacognitive strategy development, and attitudes toward the subject matter studied" (Yu, 2009, p. 1129). Although the benefits of student question generation (SQG) are well documented, a significant proportion of students do not receive explicit instruction on how to develop high-level questions, as standard STEM education typically revolves around questions of importance and of interest to the teacher (Cellitti & Wright, 2019; Minigan et al., 2017; Yu, 2009). Further, it is often assumed that students are able to generate questions without explicit teaching and repeated practice, and, as a result, teachers often attempt to dive right into implementing inquiry-based learning, overlooking this gap in ability (Cellitti & Wright, 2019). For English learners, difficulties related to proper grammatical construction of *wh*-questions add an additional layer of complexity due to the necessary inversion of the sentence order to place the *wh*-word at the beginning of the sentence, and the additional complexity of irregular *wh*-questions, such as questions that begin with *what if* or *what about*? (Cowan, 2008; Ndimele, 2015). This inversion, as well as the insertion of *do* are more complex skills for English learners, which are developed during a later stage of the question development sequence (Cowan, 2008).

To support instructors' instructional needs with regard to teaching students to develop their own questions, the use of an AI platform called SMILE.ASK (https://always.seedsofempowerment.org) was developed at Stanford University. The Stanford-Mobile-Inquiry-Based Learning Environment (SMILE) uses OpenAI's GPT-3.5/4.0 to foster critical thinking and inquiry-based learning (Kim et al., 2025). Specifically, it helps users generate, assess, and reflect on questions by providing instant feedback based on Bloom's taxonomy (Bloom, 1956). In SMILE, Bloom's six cognitive categories – Knowledge, Comprehension, Application, Analysis, Synthesis, and Evaluation – are mapped onto a five-level scale (Level 1 to Level 5), with Synthesis and Evaluation combined into Level 5, reflecting increasingly complex, higher-order thinking (HOT) as the levels increase (Hsu & Kim, 2016; Kim et al., 2025). SMILE.ASK also supports real-time monitoring, peer evaluation, and interactive discussions, thereby creating a more dynamic, student-centered classroom environment (Phan et al., 2025). In addition, for English learners, the platform offers grammatical feedback and suggestions to support the construction of *wh*-questions, including irregular forms such as *what if* questions, which are often classified at higher cognitive levels in Bloom's taxonomy.

In short, incorporating tools like SMILE-ASK into classroom instruction enables students to become more independent, thoughtful inquirers, while at the same time helping teachers foster a culture of curiosity and deeper scientific thinking. Additionally, the platform allows users to type questions in their native language, which are then automatically translated into English. A newer addition to the website allows teachers to develop a private or public classroom to collect student questions and allows for peer feedback and/or teacher feedback.

Practice 2: Developing and Using Models

This SEP coincides with Phases 2–4 of the revised inquiry process, as models can be developed as part of an Investigation (Phase 2) for use in data interpretation to figure out how different variables interact and to develop explanations; as part of the Conclusion process (Phase 3) to aid reasoning or help find relationships; and as part of the Discussion process (Phase 4) to help scientists reason with evidence and communicate results. Ultimately, models play a defining role in research questions and sharing the way data are interpreted.

Cartier et al. (2001) defined a scientific model as "a set of *ideas* that describes a natural *process*" (p. 2) noting that explanatory models go beyond physical representations, maps, diagrams, or mathematical representations to be used to explain and predict phenomena, and then revised or discarded, as new evidence is presented. Modeling begins in the K-2 grade band with pictures, dioramas, storyboards, dramatizations, and physical scale models. In the 3–5 grade band, it progresses to analogies and abstract representations, as well as diagrams of processes, which undergo revisions as more evidence is introduced. This progression continues to the use of models to predict, describe and test phenomena (including unobservable mechanisms) in the middle school grade band; and, finally, building on the K-8 experiences, to the use of models to synthesize, predict, and show relationships among variables in systems in the high school grade band (including the use of mathematical and computational models to support explanations) (NGSS Lead States, 2013, p. 53).

Further, NGSS Lead States (2013) require students in Grade 3 and above to identify the limitations of models and to revise models based on evidence to show relationships among variables. AI tools, such as ChatGPT (https://chat.openai.com/) and DALL-E (https://openai.com/dall-e), can support this SEP by enabling students to create or simulate *models of* phenomenon, and easily modify them with follow-up prompts (Additional AI tools are listed at the end of this chapter for consideration). Models developed by AI can be compared to textual explanations for additional evidence, allowing students to analyze the results of AI-generated models and use their critical thinking skills to evaluate limitations of the model, and accuracy.

Practice 4: Analyzing and Interpreting Data

This SEP coincides with Phase 2 of the inquiry process: Investigation. While authentic data collection is crucial to help students uncover patterns, identify relationships, and develop evidence-based explanations, it is often constrained by limited classroom time and resources, safety, and accessibility, which can hinder students' ability to fully engage in the Claim-Evidence-Reasoning (CER) framework (McNeill & Krajcik, 2012). In particular, students may struggle to gather robust evidence and construct sound reasoning, based on the citation of data, to support their claims. Specifically, data must be organized using tables or graphs, and undergo statistical analyses to decipher its meaning and its relevance (NGSS Lead States, 2013). Additionally, teachers may struggle to find novel datasets appropriate for various topics of study dictated in standard-based curricula.

AI tools can play a transformative role in overcoming these limitations as they enable students and teachers alike to access or generate realistic datasets, run simulations, and visualize scientific phenomena that would otherwise be inaccessible. By broadening students' access to data and analysis tools, teachers can support more equitable engagement in scientific reasoning. This not only deepens content understanding but also promotes data literacy and critical thinking, which are essential skills for academic and career success in IR4, where information and patterns may be hidden within large amounts of data (Nel & Kayembe, 2019). Finally, to support instructors' needs in terms of providing practice in data analysis, ChatGPT can be used to generate novel datasets relevant to local issues or student interests. Refinement of prompting language, details included in datasets, and formatting tips can make this process more efficient and relevant.

Literature Review

Types of Questions and Methods for Encouraging and Scaffolding Student-Generated Questions

As mentioned, aligned with SEP 1, student-generated questions serve as a cornerstone for inquiry-based science education and are critical to the NGSS's vision for K-12 science education. Chin and Osborne (2008) identified four key reasons why student-generated questions are valuable in science education. For students, (a) they guide learning and offer insight into how to link ideas; (b) they stimulate discussion and debate; (c) they support self-assessment and monitoring of understanding; and (d) they boost motivation and engagement by sparking curiosity. From a teacher's perspective, student questions provide feedback on the level of knowledge each student has attained or common misconceptions in the subject matter and, therefore, can help guide the instructional sequence (Zrada et al., 2019). Finally, asking questions and prioritizing questions are valuable skills for researchers, and are habits of mind for natural, lifelong learners in all career paths (Zrada et al., 2019).

Despite studies showing positive results from SQG, teachers are hesitant to explicitly teach and make use of instructional time to practice this skill. Besides, teachers often assume students already know how to ask insightful, testable questions and are willing to ask them (Morris & Chi, 2020). In reality, however, while preschool and primary-level students tend to find it easy to ask questions, older students often struggle to ask high-quality questions (Minigan et al., 2017). Reasons for this reluctance among older students to ask questions include social factors (fear of not knowing the answer); cognitive challenges related to developing questions (particularly for multilingual learners [MLs]); as well as instructional factors, such as a limited amount of instructional time allotted to science instruction, which may result in streamlining lessons and omitting sessions on student-generated questioning. Furthermore, in some classrooms, student questioning is discouraged, perceived as a challenge to authority, especially when teachers are under pressure to cover content quickly and adhere to mandated pacing guides. "For engineering, they [students] should ask questions to define the problem to be solved and to elicit ideas that lead to the constraints and specifications for its solution" (National Research Council, 2012, p. 560). Again, students are often not given the opportunity to define their own engineering problems or identify possible criteria and constraints related to the problem at hand (Cellitti & Wright, 2019).

For a number of reasons, therefore, in many classrooms, regular opportunities for guided practice in student questioning/problem generation are rare, and students frequently struggle to generate meaningful questions when given the opportunity and are hesitant to share their questions out loud in classroom settings. To counteract these tendencies, Zrada et al. (2019) noted that simply rephrasing the commonly used question "Do you have any questions?" to "What questions do you have?" or "Ask me two questions" encourages students to share questions, as it sets the expectation that the instructor is anticipating and welcoming student questions. Additionally, providing a turn-and-talk protocol can help students practice their questions with a peer before sharing with the class (Zrada et al., 2019).

Encouraging students to ask different types of questions is beneficial at all stages of a student's academic journey. Shakurnia et al. (2018) conducted an SQG study in an immunology course as part of postsecondary school medical education in Iran using student-generated multiple-choice questions for each chapter as a graded assignment. While most studies have shown that SQG improves student motivation (Fu-Yun & Chun-Ping, 2012), Shakurnia et al. (2018) found no overall effect on student motivation from a self-reported perspective; however, students in their SQG experimental group showed significantly higher achievement than the control group. Furthermore, low-performing students demonstrated positive perceptions toward the SQG activity, indicating that perhaps the activity resulted in lower-performing students spending more time critically analyzing and digesting the material in order to do the question development assignment (Shakurnia et al., 2018).

Given the importance of SQG, promising AI tools have been developed to help support students in overcoming these limitations by offering (a) feedback on the quality of questions asked, (b) individual question scaffolding to help correct grammatical errors, and (c) suggestions for how to improve the quality of questions developed. For example, using AI tools, students can easily generate multiple-choice questions and revise the distractors, as needed, after researching common misconceptions about the topic at hand. In such instances, AI is used to help brainstorm and draft questions, which are then further refined. Students can use detailed prompts to ensure certain vocabulary words in the subject matter are included. (A note of caution: Students will need to ensure a variety of correct answer letters exist, as some generative AI tools commonly select the same letter as a correct answer.)

Students can also use AI to scaffold the development of open-ended inquiry questions. The quality of these questions may be assessed using the well-established cognitive categories of Bloom's taxonomy. According to Yen and Halili (2015), most teacher questions rank low on Bloom's taxonomy and fail to focus on HOT. Modern versions of Bloom's taxonomy include both actions and products, for example, synthesis verbs for Level 5 include *compile*, *create*, *develop*, *generalize*, *integrate*, and *propose*, while evaluation verbs for Level 6 include *appraise*, *assess*, *criticize*, *defend*, *evaluate*, *justify*, *and support*. Although the purpose is to help students develop higher-order questions, educators should aim to honor questions at all levels on Bloom's taxonomy, as students enter the classroom with different backgrounds and different levels of understanding.

Many teachers lack explicit training in how to model and scaffold student questioning. A well-designed CPD can equip them with strategies to integrate AI tools, promote HOT, and foster a culture of curiosity, an essential component of inquiry-based learning.

Importance of Developing and Using Models in Science Classrooms

Aligned with SEP 2, modeling works hand-in-hand with developing explanations, and is used in conjunction with other SEPs. For example, analyzing data or communicating findings helps inform how models are developed, while asking questions or engaging in argument may be inspired by the ideas scientists propose in their models (Windschitl et al., 2018).

Scientific models are often used as representations of systems or phenomena, which typically include selected features of importance for understanding the coinciding system or phenomenon (Windschitl et al., 2018). However, ambiguity exists around the term *model* in science education because some interpretations of the term "tend to treat models as representations of what is known rather than as tools to be used in generating new knowledge" (Gouvea & Passmore, 2017, p. 50). For example, making a model of DNA that shows its component parts is a representation of what is already known, whereas tasking students with constructing a molecule with criteria that include the capability to copy information as cells divide and the capability to fold it, to fit inside the nucleus serves a deeper knowledge-building purpose beyond a mere depiction (Gouvea & Passmore, 2017).

Cartier et al. (2001) noted that the term *model* is used broadly, resulting in the misconception that a model is a physical replica of objects, systems, or mathematical formulae. While these may be prerequisites to the development of scientific models, the representations themselves are not explanatory models. Gouvea and Passmore (2017) agreed that scientific models are used as "tools for inquiring, not as representations of what is known" (p. 51). That is, they are the means to explain, predict, and make connections to the driving questions of the phenomenon under investigation. Oh and Oh (2011) concurred that the purpose of a scientific model is "describing, explaining, and predicting natural phenomena and communicating scientific ideas to others" (p. 1116). Thus, reproducing a picture of the water cycle, or rock cycle, or the stages of mitosis, or making a DNA molecule from pipe cleaners qualify as *models of* known scientific ideas. Similarly, graphs and charts made from data collected during a lab that show models of data "are *models of data* rather than *models of phenomena [models for]*" (Windschitl et al., 2018, p. 117). Although such representations may be useful tools, they are not the types of models that can be tested, evaluated, and revised over time (Windschitl et al., 2018).

In short, in contrast to models of data (*models of*), models of scientific phenomena or explanatory models (*models for*) are representative of an event or process, which can be used to foster critical thinking skills for students, as they evaluate new evidence and make revisions. Therefore, *models for* should explain content-rich phenomena that may be puzzling, or interesting to students, such as a singer shattering glass in an investigation of waves and resonance, or an oil tanker imploding in an investigation of atmospheric pressure and gas laws. They should include both observable and unobservable features of a phenomenon, such as body temperature, and microscopic views of the brain/muscle/skin for the runner with heat stroke phenomenon. Finally, *models for* are often used to analyze these unobservable features at different points in time by adding annotations.

SEP 2: Developing and Using Models helps students make sense of how real-world events and systems work. In doing so, working with peers, they move from using models to express initial ideas to refining them with evidence from observations, readings, and experiments. This process enables students to visualize and understand complex systems, similar to how scientists explore the natural world (Park et al., 2019). If a teacher asks students to develop a physical replica of a science phenomenon, this would be considered a *model of* and may be the starting point for further research and exploration; for example, tasking students with building a model of the layers of a plant's leaf tissue. On the other hand, using the model to predict what would happen if the tissue were placed into a salt solution (hypertonic solution) or distilled water (hypotonic solution) elevates the assignment to a *model for* phenomenon.

In summary, explanatory models (*models of*) are tentative constructions that can be assessed based on the following three criteria: (a) Can the model explain patterns in the data? (b) Can the model correctly predict the results of new experiments or observations? and (c) Is the model consistent with other ideas or other accepted scientific models? (Cartier et al., 2001). Models that do not satisfy some or all of these assessment criteria should be revised or, if necessary, discarded.

The inclusion of the following reflection questions encourages students to use a model for an epistemic purpose, which might be included in a research paper and/or a presentation to the class:

- Describe the structure and function of the parts of the system represented in your model. How does structure dictate function? What is the relationship between the scale of your model and real life? Provide a legend or key for your model.
- Describe the baseline condition of your model.
- Predict what will happen to the parts of your model with a change in the environment. Explain the cause and effect.
- Reflect: Does your model show the necessary parts to explain how the phenomenon happens? Does your model help justify your ideas to your audience? What are the limitations of your model?
- What are common misconceptions about this topic? How does your model address these misconceptions?

Studies show that *models for* aid learning even for younger students. The Learning by Design (LBD) approach (Kolodner et al., 1998) offers a pedagogical strategy for implementing SEP 2, showing that engaging students in design and modeling fosters understanding of complex systems. For example, Hmelo et al. (2000) found that Grade 6 children who created artificial lungs and built working models of select parts of the respiratory system gained deeper insight into the structure-behavior-function relationship of parts of this complex system than students who received direct instruction.

Code.org has developed *models for* lesson plans that allow students in Grades 3–12 to participate in training a bot using a dataset provided, which is used to make predictions. For example, in "AI for Oceans," at https://code.org/en-US/artificial-intelligence, the user trains a bot to sort garbage vs. sea life in the ocean. In the course of the lesson, the user tests the initial model and revises it to improve the accuracy of its predictions.

In an effort to study science teachers' experiences and views on the integration of machine learning (a subdivision of AI) into science lessons, Park et al. (2023) piloted a Grade 7 lesson package in Singapore and found that students used the machine learning model they developed to make scientific predictions. However, they also recognized the need for teacher professional development to foster confidence in incorporating them into subjects.

Analyzing and Interpreting Data in Science Classrooms

Aligned with SEP 4, although the different fields of science vary in their theory-development process, a commitment to data and evidence is a main pillar of inquiry-based learning and problem-solving in STEM subjects (National Research Council, 2012). For example, in the primary grades, explanations of phenomena are focused on description, whereas upperclassmen are required to provide explanations based on analysis, often in the form of scientific argumentation. Scientists, in turn, are required to examine, review, and evaluate their ideas, as well as those of their peers, and to use data as evidence to support their theories.

In order to carry out these complex tasks, students often need scaffolding. Scaffolds come in various formats, such as prompts, feedback, guidance, hints, and the provision of expert models (Wood et al., 1976). For example, Kim and Kim (2022) proposed that AI has the potential to help students write academic arguments by providing personalized scaffolding, which may include comparing their arguments to computer-generated samples that human experts would write. However, it is important to determine whether students simply copy and paste the AI expert suggestions or whether they are using them to internalize the corrections and improve their academic argumentation skills (Kim & Kim, 2022).

AI tools such as ChatGPT can generate and summarize data on real-world phenomena (e.g., pollution rates, population growth, carbon dioxide emission trends by country, and disease transmission). Using these resources, students can then be tasked with forming claims and supporting them with reasoning while critically evaluating the AI-generated data. That is, data generated by AI tools can be helpful for implementing the CER model in science classes, as they can provide evidence when experiments are not possible due to cost, safety, geographical barriers, or time. Besides, they offer large-scale or long-term data (e.g., climate patterns, daylight hours by month and by country) beyond what can be captured in class and allow for scenario testing and variable manipulation for deeper reasoning.

AI Literacy

Digital Promise (2024) has outlined a comprehensive approach to developing AI literacy by equipping educators, learners, and community members with the critical skills and knowledge necessary to understand, evaluate, and use AI tools responsibly and effectively. At the core of the framework are three interconnected modes of engagement: (a) Understanding, (b) Evaluating, and (c) Using AI. First, to understand AI, users must learn how it functions, how it is trained, and how it is applied across different sectors, in this context, education. This includes learning about the history of AI, machine learning processes, data usage, and the development of models like neural networks. Second, proper use of AI involves evaluating the outputs and implications of AI systems, with a strong focus on recognizing algorithmic bias, data ethics, and the societal impacts of automated decision-making. Finally, using AI refers to the application of AI tools in various forms, interacting with, creating with, and solving problems through AI, all while applying ethical judgment and ensuring data privacy and accuracy.

To operationalize these components, Digital Promise (2024) identifies three types of AI use: (a) interacting with AI (e.g., adaptive learning platforms, virtual assistants); (b) creating with AI (e.g., using generative AI to produce content); and (c) problem-solving with AI (e.g., building solutions to real-world challenges using AI). For each type, specific risks are identified, such as privacy violations or the perpetuation of bias, and corresponding AI literacy practices are recommended to mitigate these risks.

In this chapter, three types of AI applications are explored in the context of teaching science to STEM educators: question generation, data generation and analysis, and modeling. The first example demonstrates the application of the SMILE-ASK platform to support learning and instruction, a core element of AI literacy. SMILE-ASK's AI-driven question evaluation helps learners identify and refine high-quality questions, which is essential to inquiry-based learning as well as real-world problem-solving scenarios. In this context, learners use AI as a thinking partner to enhance the quality of their inquiry, an important step toward addressing meaningful problems and developing deeper understanding. The second example illustrates how AI tools can be used to collect/generate and organize data as a means of providing evidence to support scientific claims. The third example provides ways for teachers to present a phenomenon, or for students to illustrate their understanding of a scientific principle or phenomenon. Together, these examples highlight the potential of AI tools to enhance both inquiry-based pedagogy and evidence-based reasoning in STEM education.

Guiding Frameworks for Online CPD Design: Technological Pedagogical Content Knowledge and Community of Practice

Two underlying frameworks, the Technological Pedagogical Content Knowledge (TPACK) framework and Community of Practice (CoP), were leveraged for optimal support in designing online CPD for applying AI in addressing the NGSS SEPs for STEM teachers.

Technological Pedagogical Content Knowledge (TPACK)

The TPACK framework (Mishra & Koehler, 2006; see Figure 8.2) informs instructional decision-making to support teachers in thoughtfully integrating technology into their instructional practices. It is grounded in the understanding that effective technology-enhanced teaching requires integration of three core knowledge domains: *content knowledge*

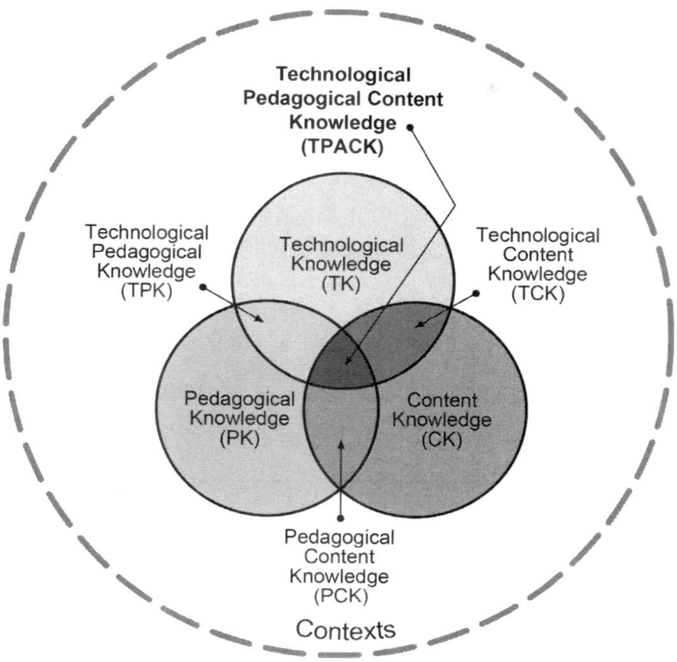

FIGURE 8.2 Technological pedagogical content knowledge

(what is being taught), *pedagogical knowledge* (how it is taught), and *technological knowledge* (the tools used for teaching). While each domain is important on its own, the true strength of the TPACK framework lies in the intersection of these areas, where meaningful and effective technology integration occurs (Mishra & Koehler, 2006). For example, a science teacher using AI tools in the classroom must understand the scientific concepts being taught (content knowledge), the best methods to help students learn through inquiry or experimentation (pedagogical knowledge), and how AI tools can support that learning (technological knowledge). When these three forms of knowledge come together, teachers are better able to design learning experiences that are meaningful, engaging, and effective.

The TPACK has been shown to be an effective framework for ensuring that technology integration in classrooms is purposeful and enhances learning outcomes. For example, Bos (2011) investigated how the TPACK framework can guide elementary teachers in integrating technology into mathematics instruction. The study involved practicing teachers designing instructional units that combined Web 2.0 tools with mathematical content. Key findings indicated that when teachers applied the TPACK framework with attention to pedagogical, mathematical, and cognitive fidelity, they were more aware of effective technology integration strategies, thereby helping to avoid misuse of instructional technology and promote the development of conceptual understanding in students. By viewing teachers as curriculum designers, the study highlighted the potential of TPACK to enhance the quality of mathematics instruction through meaningful technology integration.

In another study, Jaipal-Jamani and Figg (2015) explored how the TPACK framework can guide effective technology integration in science teaching. Three middle school science teachers participated in a four-week CPD program where they used blogs to teach the topic of fluids. The study found that authentic, hands-on experiences improved teachers' confidence and ability to integrate technology meaningfully. The authors concluded that TPACK-based PD grounded in classroom practice is effective in helping teachers design technology-enhanced science instruction.

Concurring, Mishra et al. (2023) argued that to harness AI tools effectively, educators need to develop a nuanced understanding of the interplay between technology, pedagogy, and content knowledge. This includes recognizing both the affordances and limitations of generative AI tools, designing appropriate instructional strategies, and aligning them with curricular goals. By applying the TPACK framework, educators can make informed decisions about incorporating generative AI tools in ways that enhance student learning process and outcomes.

Particularly in relation to the three SEPs (1, 2, and 4) discussed in this chapter, the TPACK framework offers a valuable foundation for designing instruction that effectively integrates AI tools into science education. That is, it helps educators and instructional designers understand the dynamic interplay among technology, pedagogy, and content knowledge, encouraging them to move beyond viewing technology as a supplemental tool and instead integrating it purposefully to support inquiry-based learning and curricular objectives (Mishra & Koehler, 2006).

Community of Practice

The second underlying framework, Communities of Practice (CoP), is defined as "groups of people who share a concern or a passion for something they do and learn how to do it better as they interact regularly" (Wenger-Trayner & Wenger-Trayner, 2015, p. 2). While knowledge communities can take many forms, CoP are typically characterized by a degree of informality and high connectivity. CoP relies on situated theories of knowledge; that is, the idea that knowledge is a property enacted by groups of people over time in shared practices, rather than the idea that knowledge is a cognitive residue in the head of an individual learner (Brown et al., 1989; Wenger, 1998).

CoP consists of the following three main characteristics: domain, community, and practice. According to Wenger-Trayner and Wenger-Trayner (2015), domain refers to a shared interest that members are committed to and in which they develop collective competence. In terms of community, members of a CoP engage in shared activities, discussions, and knowledge exchange, building relationships that foster learning. Lastly, a CoP goes beyond shared interests. Members are practitioners who build a shared repertoire of experiences, tools, and solutions through sustained interaction. Unlike a brief exchange with a stranger, this shared practice develops over time. CoPs can be a catalyst for "expansive learning," in which "… learners learn something that is not yet there. In other words, the learners construct a new object and concept for their collective activity, and implement this new object and concept in practice" (Engeström & Sannino, 2010, p. 2).

To sustain learning beyond individual modules in online CPD, the concept of CoP can be leveraged as a framework for collaborative teacher growth through small, low-stakes activities that allow learners to observe and engage with the group's routines and shared knowledge, while also leveraging technologies that support connections and conversations to facilitate their efforts, as opposed to group members trying to learn and grow in isolation.

Participating teachers can learn strategies and apply AI tools from the lens of the TPACK framework, as they have opportunities to practice and implement them in their classrooms.

Domain: The online CPD establishes a shared domain focused on integrating AI tools to teach science to address SEPs. This common focus gives participants a sense of purpose and identity as members of a professional learning community.

Community: Teachers actively engage in joint activities, such as collaborative lesson planning, online discussions, and peer feedback sessions. Through these interactions, participants share information, offer support, and build relationships that foster mutual learning.

Practice: Over time, participants develop a shared repertoire of resources, including lesson plans, AI toolkits, and strategies for implementing SEPs in science instruction by leveraging AI tools. This collective knowledge base evolves through sustained interaction, reflection, and iterative practice.

By integrating AI-enhanced science instruction grounded in the TPACK framework with the principles of CoP, the online CPD presented here equips teachers with both conceptual knowledge and practical strategies for integrating technology, purposefully considering content and pedagogy, while also fostering a collaborative environment that supports sustained professional growth.

Incorporating AI Tools in Online CPD for STEM Educators Through the Lens of the TPACK and CoP

Table 8.1 outlines the topics and sequence of sessions for the PD, based on the TPACK framework. Each session will also be described in further detail below.

TABLE 8.1 Professional Development Session Outline

Sessions	*Components*	*Topics and SEP Addressed*
Session 1	• TPACK framework	The session provides fundamental knowledge for all modules on the TPACK framework and its importance in educating students for our current IR4 society. The session addresses: • TCK in general, focusing on how technology may interact with and influence content taught in STEM classrooms. • TPK in general, focusing on how technology interacts with pedagogy and influences instructional practices in STEM classrooms. • TPACK in general, emphasizing the interconnectedness of three core knowledge domains; designing technology-enhanced learning experiences that are pedagogically sound and content-appropriate.
Session 2	• AI Literacy	The session introduces participants to AI literacy and its three components: understanding AI, evaluating AI, and using AI.
Session 3	• Technological Pedagogical Knowledge (TPK) With an Example	The session demonstrates a classroom strategy for addressing SEP 1: Asking Questions (aligned with Phase 1 of the inquiry process) by showcasing the use of SMILE.ASK, an AI-powered tool that supports students in developing higher-order questions.

(Continued)

TABLE 8.1 (Continued)

Sessions	*Components*	*Topics and SEP Addressed*
Session 4	• Technological Content Knowledge (TCK) With an Example • Technological Pedagogical Content Knowledge (TPACK) With an Example	The session demonstrates a strategy for addressing SEP 4: Analyzing and Interpreting Data by using publicly available real-world datasets (aligned with Phase 2 of the inquiry process). The session also illustrates how an AI tool, such as ChatGPT, can be used to obtain real-time data (e.g., the average monthly hours of sunlight across different countries) and to compare and contrast data from countries in the southern and northern hemispheres to help students make sense of seasonal patterns around the globe.
Session 5	• Technological Pedagogical Content Knowledge (TPACK) With an Example	The session demonstrates strategies for addressing SEP 2: Developing and Using Models (aligned with Phases 2–4 of the inquiry process). The session illustrates how an AI tool, such as ChatGPT can be used to address the topic of homeostasis in body systems by obtaining data on a runner at different stages of heat stroke to model the different physiological effects of this phenomenon before, during, and after the event. Usage of AI to develop analogies and illustrations of students' mental models is also addressed.
Subsequent Monthly Meetings	• Continued Professional Learning Online Community (Based on the Concepts of CoP and Expansive Learning)	In order to foster a culture of continuous learning and improvement, teachers reconvene virtually to share their experiences about implementing the strategies they learned and expanded upon in their content areas by incorporating AI tools. A technology platform is used to encourage ongoing collaboration, sharing of materials, and collaborative problem-solving. Steps in expansive learning as part of a CoP are addressed: questioning, analyzing, modeling, examining the model, implementing the model, and reflecting (Engeström & Sannino, 2010, p. 7).

Exemplary Activities and Resources

Session 1: Theoretical Framework: TPACK and IR4 Society

Through either synchronous or recorded video lectures, participants are introduced to the TPACK framework by first exploring its core components: Content Knowledge (CK), Pedagogical Knowledge (PK), and Technological Knowledge (TK). The session then builds on these foundations to examine the intersections – Technological Pedagogical Knowledge (TPK), Technological Content Knowledge (TCK), and the integrated TPACK framework itself. Each component is explained along with contextualized examples relevant to classroom practice, supported by visual aids such as diagrams to enhance conceptual understanding.

After this overview, participants are divided into subject-based groups, using breakout rooms in synchronous sessions or discussion forums in asynchronous formats, to collaborate on how technology can be meaningfully integrated into their teaching. Each group is asked to develop examples that illustrate the application of each TPACK component in their content area.

Sample Questions for Discussion and Reflection

- What content am I teaching?
- What pedagogical strategies suit this content?
- What technology enhances either one or both? Explain with specific examples.

Session 2: AI Literacy: Understanding AI and Evaluating AI – General

Participants are introduced to the history of AI and the concept of generative AI by watching videos and reading selected articles. After gaining this foundational knowledge, they learn about different types of prompts. With ChatGPT, the instructor demonstrates each prompt and explains how the generative AI processes and responds to it. Following the demonstration, participants have an opportunity to try out the prompts themselves, receive feedback, and refine their approach based on the guidance provided.

The following is a list of recommended activities and articles for instructors or instructional designers to use when creating lecture notes either for synchronous or asynchronous delivery (narrated or video-based for asynchronous modality).

Activity

- To Introduce the "Machine Learning" Model: https://studio.code.org/courses/oceans/units/1/lessons/1/levels/2?no_redirect=1
- *Code.org* Video: AI: What Is Machine Learning?: https://www.youtube.com/watch?v=OeU5m6vRyCk

Articles

- The History of Artificial Intelligence: https://medium.com/@ginancemonkey/the-history-of-artificial-intelligence-506882d321b0
- A Brief History of Artificial Intelligence: On the Past, Present, and Future of Artificial Intelligence: https://doi.org/10.1177/0008125619864925
- Traditional AI vs. Generative AI: What's the Difference?: https://education.illinois.edu/about/news-events/news/article/2024/11/11/what-is-generative-ai-vs-ai
- Explained: Generative AI: https://news.mit.edu/2023/explained-generative-ai-1109
- Prompt Engineering Best Practices for ChatGPT: https://help.openai.com/en/articles/10032626-prompt-engineering-best-practices-for-chatgpt
- Getting Started With Prompts for Text-Based Generative AI Tools: https://www.huit.harvard.edu/news/ai-prompts
- Effective Prompts for AI: The Essentials: https://mitsloanedtech.mit.edu/ai/basics/effective-prompts/

Sample Questions for Discussion and Reflection

- Where have you seen or experienced artificial intelligence in your life?
- How would you define generative AI in your own words?
- Which type of prompt felt most intuitive or useful to you? Why?
- What challenges did you face when trying to get accurate and meaningful responses from the generative AI?

Session 3: Technological Pedagogical Knowledge (TPK) With a Concrete Example: Use of SMILE ASK to Scaffold Student-Generated Questioning

In this session, the instructor facilitates a discussion on the importance of encouraging students to generate meaningful questions as a way to promote inquiry-based learning in science classrooms and models an inquiry-based approach:

- Video Intro: Maeyens (2018).
- Alternative: Jigsaw the article: Cellitti and Wright (2019).

Sample Questions for Discussion and Reflection

- Please share any prior teaching experiences and related frustrations.
- What additional concerns do you have that are specific to your student population? (This question is for participants whose caseloads include IEPs, 504s, or basic skills populations.)

Subsequently, participants read the following article pertaining to ASK.SMILE:

- Kim et al. (2025).

After reading the articles, participants are asked to generate four or five questions relevant to their subject area, evaluate them using ASK.SMILE, and then revise their questions based on the feedback provided. Finally, they report back to the class – verbally in the synchronous modality or by posting to the discussion board in the asynchronous class (see Figures 8.3 and 8.4). As shown in Figure 8.5, MLs, for example, can enter questions in other languages, and ASK.SMILE will generate outputs in English with suggested improvements.

Additional Resource

- Teaching materials for the updated Bloom's taxonomy to help students choose action words to aid question development or to help choose criteria for an engineering design process project: https://www.utica.edu/academic/Assessment/new/Bloom%20tx%20revised%20combined.pdf

Session 4: Technological Content Knowledge (TCK) and Technological Pedagogical Content Knowledge (TPACK) With Examples: Use of ChatGPT to Generate Data and Foster Reasoning Skills

In this session, participants learn how to use ChatGPT to evaluate hypotheses related to seasonal temperature changes in different hemispheres and present their findings. This serves as an example of how technology can support both pedagogical practices and content learning. The two main activities making up this session are as follows.

Activity 1

- The instructor asks participants to choose a city from the following list: Milan, Houston, Hong Kong, Seoul, Anchorage, Cape Town, Hanga Roa, Auckland, Lima, Sydney, and İstanbul, ensuring that the selected cities represent both the northern and southern hemispheres across all participants. It is recommended that participants use Google Earth to locate the chosen city and report its geographical coordinates (latitude and longitude).

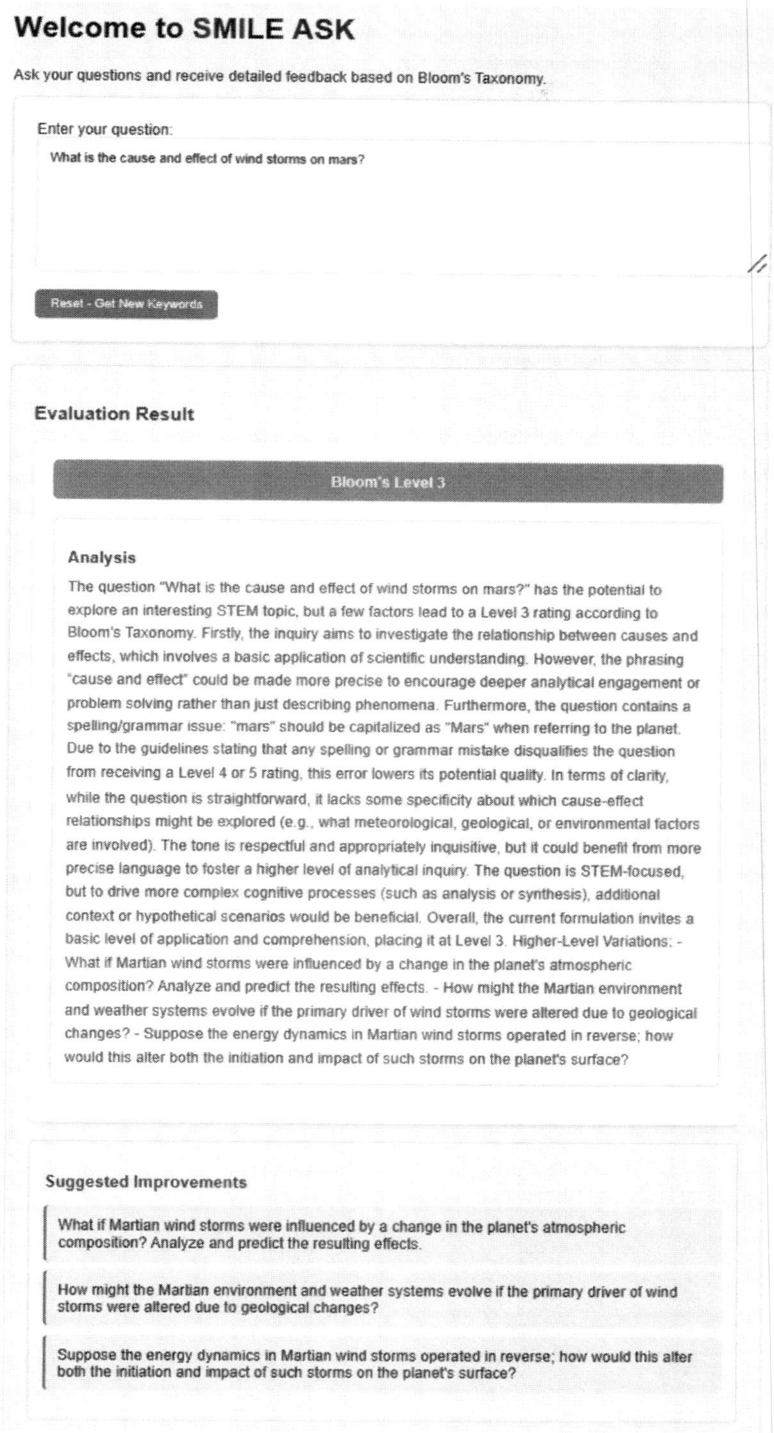

FIGURE 8.3 Input and Output Example Using ASK.SMILE for a Level 3 Question

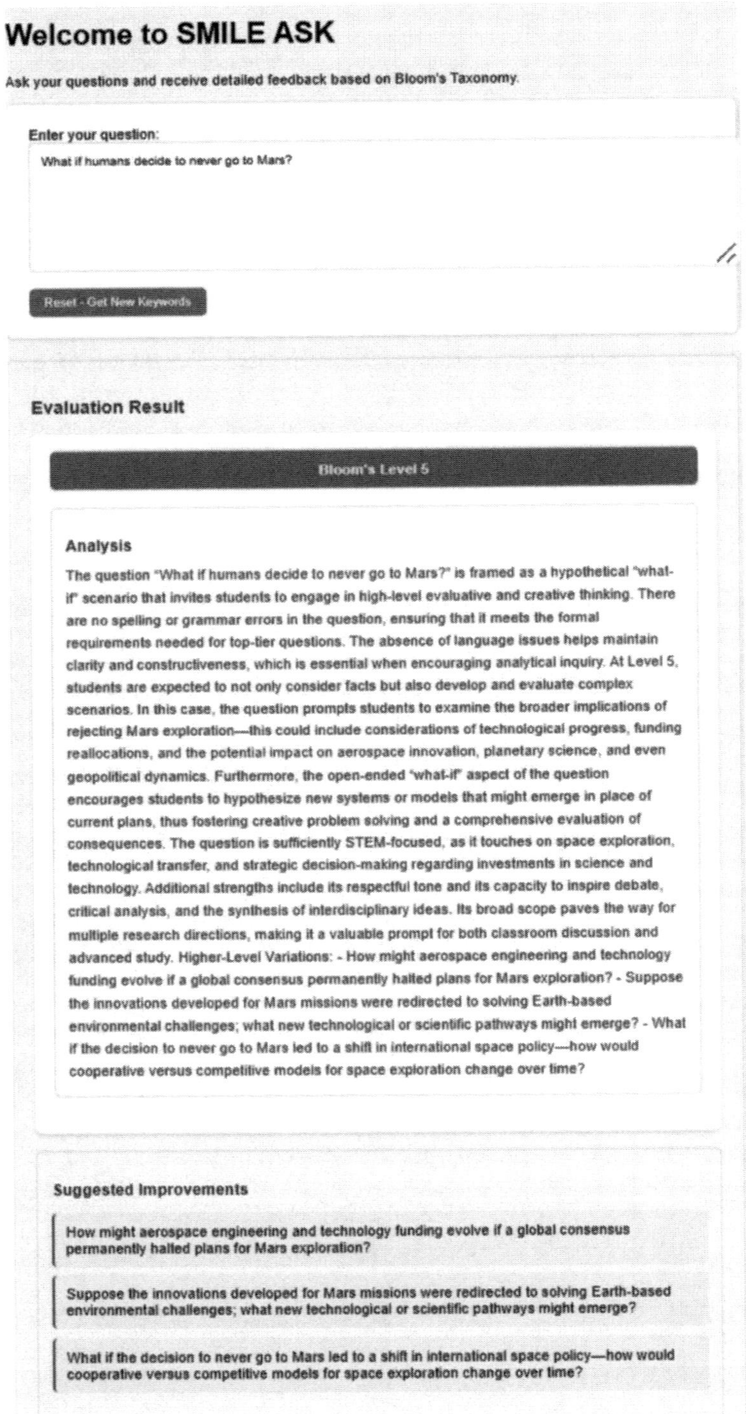

FIGURE 8.4 Input and Output Example Using ASK.SMILE for a Level 5 Question

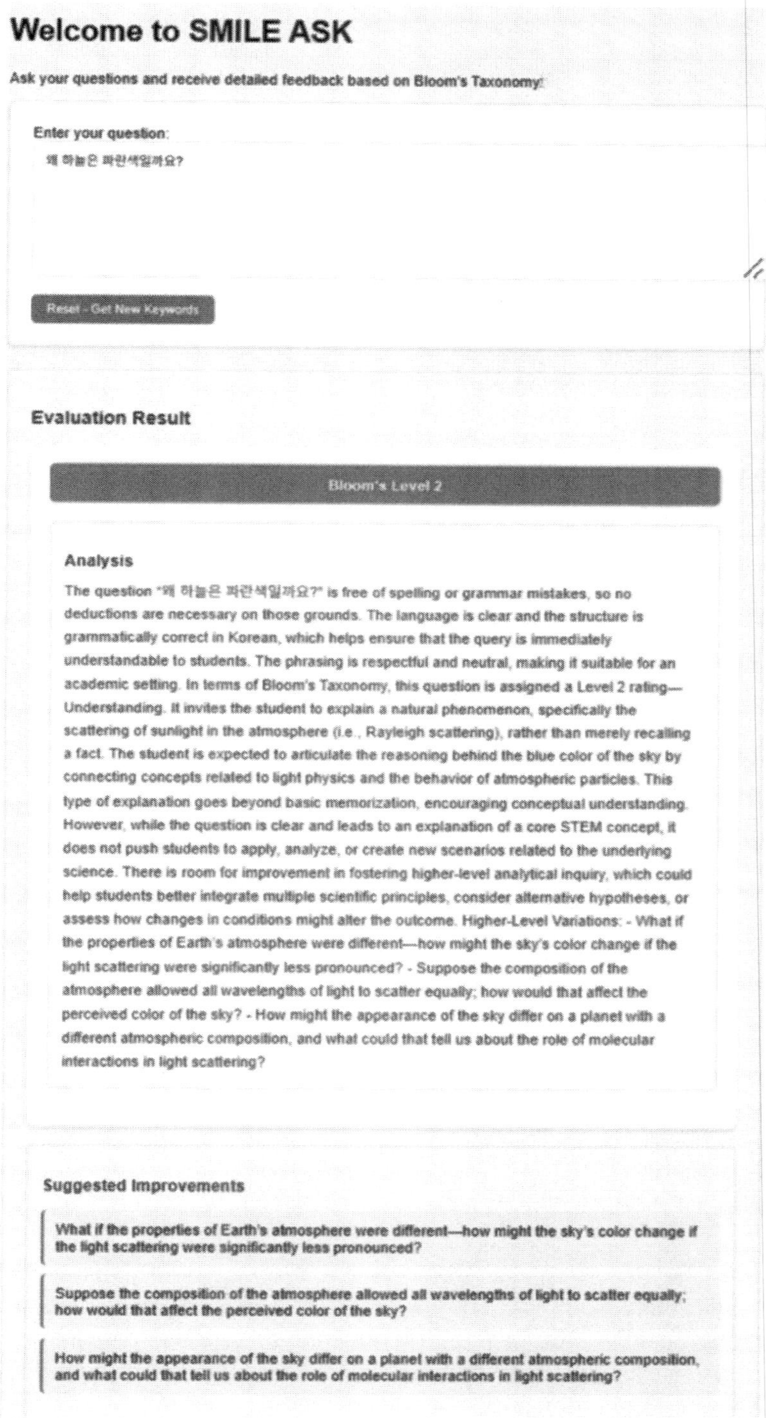

FIGURE 8.5 Example of Input (Entered in Korean) and Output (in English) Using ASK.SMILE

- Next, participants are asked to find the selected city's average monthly temperatures from the "List of Cities by Average Temperature" webpage: https://en.wikipedia.org/wiki/List_of_cities_by_average_temperature, and enter the data by month into Excel or Google Sheets and create a chart based on the data.
- The instructor demonstrates the Daylight Hours Calculator (https://jan.moesen.nu/daylight-calculator), showing participants how they can enter cities, states, or countries to further explore and compare daylight hours.

Activity 2

- The instructor presents the following question "What is the relationship between latitude and the number of daylight hours on Earth?" and asks participants to write their claims in response. Participants are to collect relevant data (evidence) using ChatGPT to support or refute their claims, following the CER model.

> Question. "What is the relationship between latitude and the number of daylight hours on earth?"
>
> Claim: ____________________
>
> Evidence: ______ [*We can collect data using an AI tool, ChatGPT*] ____________
>
> Reasoning: ____________________

Gathering Data Using ChatGPT

For synchronous online CPD, the instructor demonstrates the process step-by-step, allowing participants to follow along while selecting cities and sharing their tables and charts in the chat. For asynchronous online CPD, it is recommended that the instructor record a step-by-step tutorial so participants can follow the instructions at their own pace and then post their tables and charts on the discussion board.

The instructor asks participants to open ChatGPT and provides an example prompt to collect the data. The instructor then guides participants in gathering data for two cities: one from the northern hemisphere and one from the southern hemisphere. Once the data have been generated into a table, they may be downloaded and imported into Excel or Google Sheets to create a chart. See Figures 8.6 and 8.7.

Prompt: "Find the average daylight hours for each month in ______ and make a table."

Reasoning

For synchronous online CPD, the activities may be conducted in breakout groups, with each group reporting their completed CER in the main Zoom room. For asynchronous online CPD, the same activities can be carried out through a group discussion board, with participants sharing their findings on the main discussion board.

Participants compare and contrast their charts, discussing why some charts display a deep curve and what factors contribute to its formation. They then revisit their initial claims

Prompt: "Find the average daylight hours for each month in Seoul and Sydney and make a table."

Month	Seoul (h/day)	Sydney (h/day)
Jan	9.86	14.16
Feb	10.72	13.38
Mar	11.86	12.37
Apr	13.08	11.32
May	14.12	10.43
Jun	14.70	9.95
Jul	14.51	10.10
Aug	13.65	10.82
Sep	12.52	11.79
Oct	11.32	12.85
Nov	10.24	13.80
Dec	9.64	14.35

FIGURE 8.6 Example of a Table Using a Dataset Created With ChatGPT

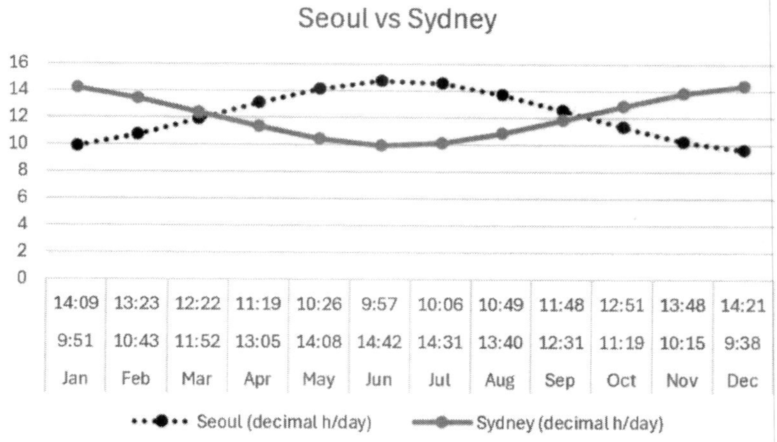

FIGURE 8.7 Example of a Figure Chart Based on the Data Shown in Table

and make modifications if necessary. Finally, participants brainstorm ideas for reasoning based on the evidence they collected (tables and figures) and articulate this reasoning in their CER notes.

Session 5: Technological Pedagogical Content Knowledge (TPACK) With an Example: Use of AI Tools to Support the Development of Scientific Models

Participants are introduced to applications of AI models through videos and articles. In Activity 2, teachers compare *models of* and *models for* in order to explore different ways models are used in science and then refer to the progressions in modeling in their particular grade bands. Activity 3 provides a demonstration of three different ways ChatGPT can be used to develop and iterate models. Participants have an opportunity to try out modeling themselves and share feedback.

Activity 1: Modern-Day Modeling Uses of AI

Three different uses of AI models are explored by reading and watching videos: (a) the use of AI modeling in a weather prediction application currently being tested by the National Oceanic and Atmospheric Administration that uses predictive modeling for severe weather patterns with increased accuracy, based on current conditions combined with historical data as opposed to the High-Resolution Rapid Refresh (HRRR) model, which is a physics-based model; (b) the use of AI modeling for predicting genetic diseases based on facial features; and (c) AI-enabled medical devices healthcare models (which incorporate an array of AI systems ranging from large language models to multimodal architectures that gather and synthesize information from different sources).

- (a) NOAA Research Develops an AI-Powered Sibling to Its Flagship Weather Model. (2025).
- (b) Video – *AI Tech Can Identify Genetic Disorders from a Person's Face*. https://mashable.com/video/ai-identify-genetic-disorders-photo?
- (c) U.S. Food and Drug Administration (2025).

Sample Questions for Discussion and Reflection

- What are the benefits and drawbacks of using generative AI in the applications viewed?
- What other applications of generative AI models currently affect our lives?

Activity 2: Jigsaw the Articles

Participants read articles (a) and (b) below, and discuss basic concepts about developing and using models, along with some common misconceptions. After reading the articles, participants then view the NGSS expectations in this SEP as students move through grade levels from K-12. These grade band progressions, discussed in article (c), exist for each SEP (NGSS Lead States, 2013). Finally, article (d) provides an example of a carbon sequestration research project that incorporates the use of AI and analyzes prompting sequences to determine which populations of students may benefit most from such a project.

(a) Oh & Oh (2011).
(b) Discussion on *models of* versus *models for*. (See Gouvea & Passmore, 2017, pp. 59–60, for examples.)
(c) Progressions in modeling across grade bands (NGSS Lead States, 2013, Volume 2, p. 53).
(d) The following study required students to develop mental models of carbon sequestration with the use of AI online information-searching strategies (Lai, 2025).

Sample Questions for Discussion and Reflection

- Please share prior teaching experiences in the development and use of models in your teaching experience, and any related frustrations.
- How have you differentiated this SEP for different segments of your student population? (This question is for participants whose student populations include MLs, IEPs, 504s, or basic skills populations.)
- What new insights have you gained in developing models?

Activity 3: Modeling Examples

Participants are introduced to ways of creating various types of models with the aid of ChatGPT. The following three examples illustrate how this can be done: (a) Use of sequential modeling – demonstrates how ChatGPT can be used to generate step-by-step representations of processes or concepts; (b) Use of ChatGPT to generate an analogy as a model – shows how ChatGPT can create meaningful analogies that simplify abstract or complex ideas by connecting them to familiar concepts; and (c) Use of AI to make mental models visible – highlights how ChatGPT can externalize learners' internal thinking, thereby making implicit understandings more explicit and open for reflection.

(a) Use of Sequential Modeling

The instructor asks participants to: (a) ask ChatGPT to model a homeostasis phenomenon of their choice before, during, and after a negative feedback loop; (b) analyze the limitations of the model; and (c) based on the model, develop a fictional story told from the first-person point of view. In the story, addresses at least two of the limitations of the ChatGPT model.

- Prompt 1: "Pretend you are a high school science teacher. Design a model of a runner before, during, and after heat stroke. Include statistics on muscles at the microscopic level, pulse, breathing rate, temperature, and blood pressure." (Response: See table presented in Figure 8.8. Additionally, ChatGPT asked, "Do you want me to make a student-friendly infographic (runner outline + arrows and vitals at each stage) so it's easy to visualize in class?".)
- Prompt 2: "Make a student-friendly infographic." (See Figure 8.9 for response.)
- Prompt 3: "Explain the limitations of this model." (See Figure 8.10 for response.)

Runner Model: Before, During, and After Heat Stroke

Phase	What's going on (big picture)	Microscopic muscle changes	Pulse (bpm)	Breathing rate (breaths/min)	Core temp (°C/°F)	Blood pressure (mmHg)
Before (Resting state)	Body is balanced, normal thermoregulation; heart and lungs at baseline.	Normal myofibril alignment; mitochondria functioning normally; intact sarcolemma.	60–100	12–20	36.5–37.5°C (97.7–99.5°F)	~120/80
During Heat Stroke (medical emergency)	**Thermoregulatory failure**: heat production > heat loss. **Core ≥40°C plus CNS dysfunction** (confusion, collapse). Marked dehydration → low plasma volume.	**Heat-induced sarcolemma disruption, myofibrillar disarray** (**Z-line streaming**), **mitochondrial swelling** → muscle cell injury.	Often >120	Often >20–30	**≥40°C (104°F)**; may reach 40.5–42°C	**Drops to ~90/60 or lower** due to dehydration + vasodilation
After (Recovery with treatment)	Cooling and fluids restore balance; organs monitored for complications.	Damaged fibers cleared by immune cells; repair and regeneration begin.	Returns toward baseline (often 80–110 soon after).	Returns toward **12–20**	Falls back below **39°C** with cooling	Gradual normalization, often **110–120/70–80**

FIGURE 8.8 AI-Generated Image

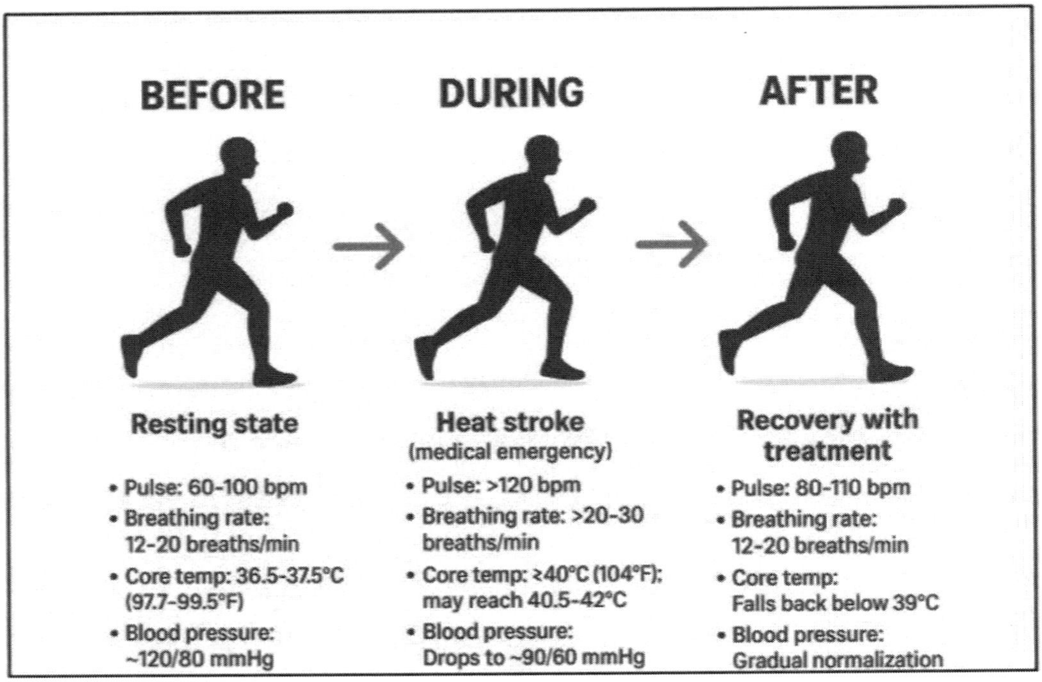

FIGURE 8.9 AI-Generated Image

(b) Use of ChatGPT to Generate an Analogy as a Model

Models can be made in the form of analogies to help students make connections to prior knowledge or to everyday activities. Participants are asked to do the following activity: "Ask ChatGPT to develop an analogy for a phenomenon in your subject area and level. Explain the similarities and limitations of the model. Then, refine the model." The following are two examples.

Here are some key **limitations of the runner heat stroke model** you saw in the infographic:

1. **Simplified physiology**
   - Real responses vary with age, fitness, hydration, and environmental conditions. The vitals listed are typical rangers, not universal values.
2. **Focus on vitals only**
   - The model highlights pulse, breathing rate, temperature, and blood pressure, but doesn't capture other critical systems affected (like brain function, kidney injury, electrolyte disturbances, or clotting disorders).
3. **Microscopic muscle changes generalized**
   - While heat can damage muscle fibers and mitochondria, the exact degree of cellular injury differs across individuals and isn't visible in real time like the infographic suggests.
4. **No overlap of phases**
   - The infographic shows "before, during, after" as neat stages, but in reality, there can be a continuum: mild heat exhaustion may precede heat stroke, and recovery may be incomplete or prolonged.
5. **Treatment oversimplified**
   - The "after" stage assumes rapid and successful medical intervention. Without immediate and aggressive cooling, outcomes can be severe, including multi-organ failure.
6. **Context missing**
   - Environmental factors (humidity, direct sun, clothing, hydration, fitness level) are critical in determining whether heat stroke occurs but are not shown.

FIGURE 8.10 AI-Generated Image

Analogy Example 1: Life Science

- Prompt 1: "Pretend you are a teacher in a high school biology class. Develop and explain an analogy for cellular respiration. Include the similarities to real life and limitations of the model." Participants critique and refine this analogy based on textual evidence (See Figure 8.11 for response.)
- Prompt 2: "What misconceptions might students have about this topic?" (See Figure 8.12 for response.)

Analogy Example 2: Physical Science

- Prompt 1: "Pretend you are a high school physics teacher, provide an analogy for different types of electrical circuits." (See Figure 8.13 for response. Additionally, ChatGPT asked, "Would you like me to also make this analogy into a student-friendly diagram (like a cartoon of a water park showing series vs parallel slides)?".) Participants critique and refine the analogy response based on textual evidence. An expected response from participants might note that in a circuit electrons run in a loop to close the circuit; however, a water slide typically runs downhill requiring riders to walk back up the stairs to complete the circuit. If the riders do not walk back up to complete the circuit, there will

Analogy: The School Cafeteria as the Cell's Power Plant

Imagine your school cafeteria as the cell, and food as glucose. Just like students need food to get energy for the day, cells need glucose to make usable energy in the form of **ATP.**

1. **Cafeteria receives ingredients (Glucose + Oxygen)**
- Delivery trucks bring in food supplies (glucose) and oxygen tanks (O_2).
- In the cell, glucose comes from the food you eat, and oxygen comes from breathing.
2. **Kitchen Staff = Enzymes**
- The cooks (enzymes) break down the ingredients in a step-by-step recipe.
- This represents glycolysis, the Krebs cycle, and the electron transport chain.]
3. **Cooking = Breaking Bonds**
- Just like chopping, boiling, or grilling releases flavors and heat, breaking chemical bonds in glucose releases energy.
4. **Finished Meals = ATP**
- The energy released doesn't just fly away – it's captured and packaged into "meals" (ATP).
- ATP is like pre-packed lunches, small and usable for students (the cell's activities).
5. **Waste Disposal**
- Leftover trash from cooking = carbon dioxide (CO_2)
- Dirty dishwater = water (H_2O)
- Just like the cafeteria has waste, cells produce CO_2 and H_2O as byproducts.

Similarities to Real Life

- Food in cafeteria = glucose for cells
- Cooks = enzymes guiding each step
- Cooking releases usable energy = glucose breakdown releasing ATP
- Trash and dishwater = carbon dioxide and water
- Meals fueling students = ATP fueling cellular work (muscle contraction, nerve signals, growth)

Limitations of the Model

- In real life, cafeterias don't turn food into energy directly – they make meals, not chemical fuel.
- Cooking is not as precise or controlled as enzyme reactions in cells.
- Students eat meals over hours, but cells use ATP almost instantly.
- Oxygen isn't delivered in tanks to kitchens, but it is essential in respiration.

FIGURE 8.11 AI-Generated Image

Common Misconceptions

1. **Glucose → ATP directly**
    - Students may think glucose instantly turns into ATP, like unwrapping a snack.
    - Reality: glucose must go through multiple steps (glycolysis, Krebs cycle, electron transport chain).
2. **ATP is stored long-term like food**
    - They may think ATP is stockpiled like meals in a fridge.
    - Reality: ATP is used almost immediately and is constantly regenerated.
3. **Oxygen is used to break down glucose directly**
    - Students might imagine oxygen "cutting up" glucose.
    - Reality: oxygen acts as the final electron acceptor in the electron transport chain.
4. **Respiration is the same as breathing**
    - Many confuse "cellular respiration" with inhaling and exhaling.
    - Breathing delivers oxygen and removes CO_2, but the chemical reactions happen inside cells.
5. **Energy is "made" rather than transferred**
    - Students may think cells create new energy out of nothing.
    - Reality: energy from glucose's chemical bonds is transformed into ATP's chemical energy (following the law of conservation of energy).
6. **Byproducts are waste and useless**
    - They may assume CO_2 and H_2O are just trash.
    - Reality: CO_2 is crucial for plants in photosynthesis, and water is essential for many cellular processes.
7. **All organisms do photosynthesis + respiration**
    - Some believe plants onto photosynthesize, or that animals only respire.
    - Reality: both plants and animals respire; only plants, algae, and some bacteria also photosynthesize.

FIGURE 8.12 AI-Generated Image

be an open circuit and a break in the flow of electrons, and electricity, just like when a light switch is turned off.

- Prompt 2: "Yes." (See Figure 8.14 for response. Note that Figure 8.14 reinforces the possible misconception of circuits as linear shapes [open circuits] in the analogy as opposed to a loop shape needed to close the circuit.)
- Prompt 3: "Can you add to the analogy by adding resistance to the circuit?" (See Figure 8.15 for response.)

(c) Use of AI to Make Mental Models Visible: Genetics Example

AI can be used by teachers to develop a model of a phenomenon, or by students to illustrate a fantasy species that they develop and design traits to, inherited through different types of inheritance patterns. For example, AI can be used to begin with a family inheritance phenomenon. Students can be shown a family tree of fantasy character images that illustrate

Analogy: Water Park Rides

Think of an amusement park with water rides. The water represents the **electric current,** the pipes are the **wires,** and the pumps pushing the water are like the **battery or power source.**

1. **Series Circuit → One Long Waterslide**
    - Imagine a single, twisty waterslide.
    - Everyone has to go down one after another.
    - If the slide gets blocked halfway (like a burned-out bulb), nobody else can get through- just like how current stops if one part of a series circuit is broken. The riders' energy gets used up gradually along the way, just like voltage drops across each resistor.
2. **Parallel Circuit → Multiple Waterslides Side by Side**
    - Picture several waterslides starting from the same tower.
    - Riders can choose which slide to go down.
    - If one slide is closed for repairs, the others still work- like how current can still flow through other branches of a parallel circuit.
    - Each rider gets the full thrill of the ride, just as each branch gets the full battery voltage.
3. **Combination Circuit → A Tower with Branches and Loops**
    - Some water rides start as one big slide, split into two different paths, and then join back together at the bottom.
    - Riders split up and regroup, just like current dividing in parallel section and then coming back together.

Limitations of the Analogy:

- Water flow depends on gravity and pressure, but electric charges move because of electric fields.
- People on slides move in chunks (discrete), while electric current is continuous.

Would you like me to also make this analogy into a **student-friendly diagram** (like a cartoon of a water park showing series vs parallel slides)?

FIGURE 8.13 AI-Generated Image

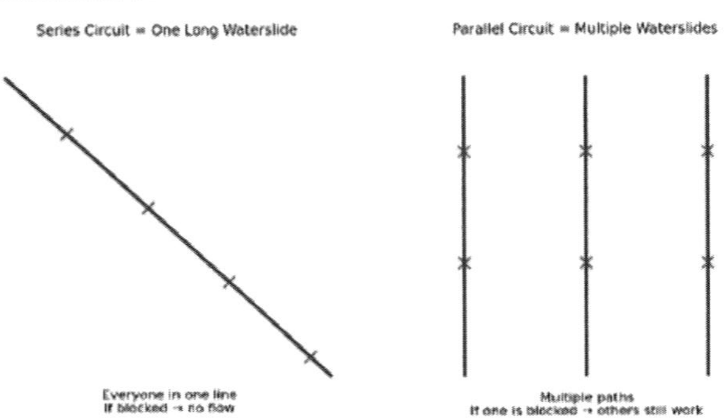

Here's a **student-friendly diagram** showing the water park analogy for circuits:

- On the left: **Series circuit** → one long waterslide where everyone must go in order.
- On the right: **Parallel circuit** → multiple waterslides side by side, each still working even if one closes.

Would you like me to also add a **combination circuit slide tower** to complete the analogy?

FIGURE 8.14 AI-Generated Response

Great question! Let's add **resistance** into our water park analogy so it clicks for students:

Resistance = Friction or Obstacles on the Waterslide

- **Low Resistance** A smooth waterslide with lots of water. Rides zoom down quickly.
- **High Resistance** A rough slide, with bumps, twists, or sponges that slow riders down.
- **Extreme Resistance** A corked slide – no riders can pass (like an open circuit).

Applying this to circuits:

- In a **series circuit,** more resistors are like adding more rough patches along the same slide – each one slows riders down more.
- In a **parallel circuit,** each slide may have its own level of roughness. Riders will naturally prefer smoother (low resistance) slide – just like electric current favors the easier path.

Limitation of the Analogy: In real circuits, current splits mathematically according to resistance (Ohm's Law), while in a water park, people just pick whichever slide looks easier.

Would you like me to **update the diagram** to show "resistance" with things like bumps, sponges, or rocks on the slides?

FIGURE 8.15 AI-Generated Response

different types of inheritance. As the unit progresses, students explain which traits are inherited through which types of inheritance. As a culminating project, students can create their own fantasy species and explain the inheritance pattern for each trait. See Figure 8.16 for a section of a possible family tree of fantasy characters.

Notably, images requested using ChatGPT do not always provide accurate results. While the prompt for requesting a drawing of Pedro showed the requested results, prompts for Delores and Chimera did not. After drawing Pedro with ChatGPT, the user requested, "Draw the same monster with pink ears, blue eyes, white skin with no spots, pink feet, and 10 arms," to alter Pedro's image; however, the resulting image only contained six arms. Therefore, the genotypes were revised to match the intended images. Furthermore, for Chimera, the request, "Draw the same monster with small pink ears, green pointed eyes, purple spiked straight hair, light blue skin, blue and pink striped feet, no antennae, two arms per side and six fingers" resulted in an image with only two arms and with light blue ears instead of pink ears. These are examples of hallucinations (i.e., false, misleading, or nonsensical AI responses), as a result of the way generative AI models are trained. After receiving a response, to improve accuracy, the user can try adding the following request, "Be sure that the response meets all the requirements from the previous prompt."

FIGURE 8.16 AI-Generated Individual Images

The instructor can ask the following discussion questions to facilitate further conversations with participants, either verbally or through a discussion board.

- How accurate were the models AI designed? What biases or hallucinations did you identify in the AI-generated responses? What type of refinements were required in order to make the image match your mental model?
- How can the prompts used be refined to obtain more accurate results from the beginning?
- What are the advantages of using AI modeling with your population of students? What issues might you face in implementing an AI modeling assignment in your classroom?
- How should we cite the use of AI used to generate text and images?
- How does AI affect human creativity?

Conclusion

As STEM education evolves to meet the demands of a rapidly changing technological landscape, integrating AI tools into inquiry-based learning presents both exciting opportunities and critical pedagogical challenges. As illustrated in this chapter, AI tools such as ASK. SMILE and ChatGPT have the potential to enhance key scientific practices, such as generating questions and gathering and analyzing data to support claims, by fostering student engagement, promoting critical thinking, and broadening access to authentic, data-driven inquiry experiences. However, effective integration of these tools requires more than basic familiarity with current technology. It also demands intentional instructional design guided by frameworks such as TPACK.

Online CPD grounded in the TPACK framework equips educators to better align technological tools with content objectives and pedagogical strategies, ensuring that AI is not used as a superficial add-on but as an integral component of student learning. Through concrete examples and structured online CPD sessions, this chapter has demonstrated how AI tools can support the development of core SEPs outlined in the NGSS, particularly in helping students generate meaningful questions and engage in evidence-based reasoning.

Equally important is the cultivation of a CoP that sustains professional learning beyond individual training sessions. A well-facilitated CoP fosters ongoing dialogue, collaborative problem-solving, and the sharing of best practices among educators. It enables teachers to continually reflect on their practice, refine their use of AI tools, and collectively navigate the evolving challenges of AI-integrated STEM instruction. By fostering both AI literacy and inquiry-based pedagogy within a supportive professional community, teachers will be able to prepare their students not only to thrive academically but also to think critically and act responsibly in their use of AI tools to solve complex, real-world problems.

Additional AI Tools

Conversational AI/Chatbots

- ChatGPT (OpenAI): https://chat.openai.com/
- Claude (Anthropic): https://claude.ai/
- Copilot (Microsoft): https://copilot.microsoft.com/
- Gemini (Google DeepMind): https://gemini.google.com/
- Perplexity AI (Perplexity AI, Inc.): https://www.perplexity.ai/
- Pi (Inflection AI): https://pi.ai/

Image Generators

- Bing Image Creator (Microsoft): https://www.bing.com/images/create
- Copilot (Microsoft): https://copilot.microsoft.com/
- DALL-E (OpenAI): https://openai.com/dall-e
- Ideogram (Ideogram AI): https://ideogram.ai/t/explore
- Leonardo AI (Leonardo Interactive Pty Ltd): https://leonardo.ai/
- NightCafe (NightCafe Studio): https://creator.nightcafe.studio/
- PicLumen (PicLumen Inc.): https://piclumen.com/
- Playground (Playground AI): https://playground.com/
- Stable Diffusion (Stability AI): https://stablediffusionweb.com/

KEY TAKEAWAYS

- TPACK offers a robust framework for aligning technology with instructional goals and content.
- AI tools like SMILE-ASK and ChatGPT can deepen inquiry-based STEM learning by enhancing students' ability to ask high-quality questions, analyze real-world data, and develop models, all of which are core practices aligned with SEPs. These tools support meaningful engagement with the CER framework, promoting critical thinking and evidence-based reasoning.
- Effective online CPD for STEM teachers must go beyond tool training. It should foster AI literacy, understanding, evaluating, and using AI tools, while also supporting pedagogical strategies for implementing inquiry-based instruction. A TPACK-informed online CPD model empowers educators to design student-centered, AI-enhanced science learning experiences that more effectively support content mastery.
- Sustained integration of AI in STEM education is best supported through a strong CoP, where teachers continuously and collaboratively reflect, share, and grow as they explore and implement AI tools in their own classrooms.
- Currently, there are limitations, biases and inaccuracies when using AI tools. Instructors need to be cognizant of hallucinations or misleading responses, and teach their learners to be critical thinkers and skeptical consumers of AI-generated information.

References

Arucu, M. (2021). Industry 4.0 – Holistic perspective: Modeling and simulation implementations in manufacturing systems. *Journal of Innovative Science and Engineering*, *5*(1), 50–63. https://doi.org/10.38088/jise.816023

Bloom, B. S. (1956). *Taxonomy of educational objectives: The classification of educational goals*. D. McKay Co., Inc.

Bos, B. (2011). Professional development for elementary teachers using TPACK. *Contemporary Issues in Technology and Teacher Education*, *11*(2). https://citejournal.org/volume-11/issue-2-11/mathematics/professional-development-for-elementary-teachers-using-tpack

Brown, J. S., Collins, A., & Duguid, P. (1989). Situated cognition and the culture of learning. *Educational Researcher*, *18*(1), 32–42.

Cartier, J., Rudolph, J., & Stewart, J. (2001). *The nature and structure of scientific models: Working paper*. ERIC (ED461513). https://eric.ed.gov/?id=ED461513#:~:text=ERIC%20Number:%20ED461513,Literacy%2C%20Scientific%20Methodology%2C%20Scientific%20Principles

Cellitti, J., & Wright, C. (2019). Encouraging student-generated questioning. *Science and Children*, *57*(4), 76–79. https://ezproxy.wpunj.edu/login?url=https://www.proquest.com/scholarly-journals/encouraging-student-generated-questioning/docview/2312779564/se-2

Chin, C., & Osborne, J. (2008). Students' questions: A potential resource for teaching and learning science. *Studies in Science Education*, *44*, 1–39. https://doi.org/10.1080/03057260701828101

Chiu, T.K.F., Xia, Q., Zhou, X., Chai, C. S., & Cheng, M. (2023). Systematic literature review on opportunities, challenges, and future research recommendations of artificial intelligence in education. *Computers and Education: Artificial Intelligence*, *4*, 100118. https://doi.org/10.1016/j.caeai.2022.100118

Cowan, R. (2008). *The teacher's grammar of English: A course book and reference guide*. Cambridge University Press.

Delello, J. A., Sung, W., Mokhtari, K., Hebert, J., Bronson, A., & De Giuseppe, T. (2025). AI in the classroom: Insights from educators on usage, challenges, and mental health. *Education Sciences*, *15*(2), 113. https://doi.org/10.3390/educsci15020113

Digital Promise. (2024, June 18). *AI literacy: A framework to understand, evaluate, and use emerging technology*. https://digitalpromise.org/2024/06/18/ai-literacy-a-framework-to-understand-evaluate-and-use-emerging-technology/

Du, H., Sun, Y., Jiang, H., Islam, A.Y.M.A., & Gu, X. (2024). Exploring the effects of AI literacy in teacher learning: An empirical study. *Humanities & Social Sciences Communications*, *11*(1), 559. https://doi.org/10.1057/s41599-024-03101-6

Engeström, Y., & Sannino, A. (2010). Studies of expansive learning: Foundations, findings and future challenges. *Educational Research Review*, *5*(1), 1–24. https://doi.org/10.1016/j.edurev.2009.12.002

Fu-Yun, Y., & Chun-Ping, W. (2012). Student question-generation: The learning processes involved and their relationships with students' perceived value. *Journal of Research in Education Sciences*, *57*(4), 135–162. https://doi.org/10.3966/2073753X2012125704005

Gouvea, J., & Passmore, C. (2017). "Models of" versus "Models for." *Science & Education*, *26*, 49–63.

Hmelo, C. E., Holton, D. L., & Kolodner, J. L. (2000). Designing to learn about complex systems. *The Journal of the Learning Sciences*, *9*(3), 247–298. http://www.jstor.org/stable/1466843

Hsu, H. Y., & Kim, P. (2016). Preservice teachers' uses of SMILE to enact student-generated questioning 742 practices. *International Journal of Innovation in Education*, *3*(2–3), 110–121.

Jaipal-Jamani, K., & Figg, C. (2015). A case study of a TPACK-based approach to teacher professional development: Teaching science with blogs. *Contemporary Issues in Technology and Teacher Education*, *15*(2), 161–200. https://citejournal.org/volume-15/issue-2-15/science/a-case-study-of-a-tpack-based-approach-to-teacher-professional-developmentteaching-science-with-blogs

Kim, M. K., & Kim, N. J. (2022). AI supported scaffolding for writing academic arguments. In *Proceedings of the international society of the learning sciences*.

Kim, P., Wang, W., & Bonk, C. J. (2025). Generative AI as a coach to help students enhance proficiency in question formulation. *Journal of Educational Computing Research*, *63*(3), 565–586. https://doi.org/10.1177/07356331251314222

Kolodner, J. L., Crismond, D., Gray, J., Holbrook, J., & Puntambekar, S. (1998). Learning by design from theory to practice. In *Proceedings of ICLS 98* (pp. 16–22). AACE.

Lai, C.-L. (2025). Consolidate knowledge or build scientific models? The role of online information-searching strategies in students' prompt sequences with generative artificial intelligence. *Educational Technology & Society*, *28*(3), 241–262. https://doi.org/10.30191/ETS.202507_28(3).TP03

Maeyens, K. (2018, June 4). *TEDxUFM: The value of asking questions*. https://www.youtube.com/watch?v=aZIuAQw8RA4

McNeill, K. L., & Krajcik, J. (2012). *Supporting grade 5–8 students in constructing explanations in science: The claim, evidence, and reasoning framework for talk and writing*. Pearson Allyn & Bacon.

Minigan, A. P., Westbrook, S., Rothstein, D., & Santana, L. (2017). Stimulating and sustaining inquiry with students' questions. *Social Education*, *81*(5), 268–272. https://ezproxy.wpunj.edu/login?url=https://www.proquest.com/trade-journals/stimulating-sustaining-inquiry-with-students/docview/3156825591/se-2

Mishra, P., & Koehler, M. (2006). Technological pedagogical content knowledge: A framework for teacher knowledge. *Teachers College Record*, *108*(6), 1017–1054. https://doi.org/10.1111/j.1467-9620.2006.00684.x

Mishra, P., Warr, M., & Rezwana Islam, R. (2023). TPACK in the age of ChatGPT and generative AI. *Journal of Digital Learning in Teacher Education*, *39*(4), 235–251. https://doi.org/10.1080/21532974.2023.2247480

Morris, J., & Chi, M.T.H. (2020). Improving teacher questioning in science using ICAP theory. *The Journal of Educational Research*, *113*(1), 1–12. https://doi.org/10.1080/00220671.2019.1709401

National Research Council (NRC). (2000). *Inquiry and the national science education standards: A guide for teaching and learning*. National Academy Press.

National Research Council (NRC). (2012). *A framework for K-12 science education: Practices, crosscutting concepts, and core ideas*. The National Academies Press. https://nap.nationalacademies.org/download/13165#

Ndimele, O. (Ed.). (2015). *Language endangerment: Globalisation and the fate of minority languages in Nigeria*. M & J Grand Orbit Communications Limited. https://ebookcentral.proquest.com/lib/wpunj-ebooks/reader.action?pq-origsite=primo&c=UERG&ppg=1&docID=4560544

Nel, D., & Kayembe, C. (2019). Challenges and opportunities for education in the Fourth Industrial Revolution. *African Journal of Public Affairs*, *11*(3), 79–94. https://journals.co.za/doi/epdf/10.10520/EJC-19605d342e

Next Generation Science Standards. (n.d.). *Next generation science standards*. https://www.nextgenscience.org/

NGSS Lead States. (2013). *Next generation science standards: For states, by states*. The National Academies Press.

Niemi, H., Pea, D., & Yu, L. (Eds.). (2023). *AI in learning: Designing the future*. Springer International Publishing. https://doi.org/10.1007/978-3-031-09687-7

NOAA Research Develops an AI-Powered Sibling to Its Flagship Weather Model. (2025, July 23). *Targeted News Service*. https://research.noaa.gov/noaa-research-develops-an-ai-powered-sibling-to-its-flagship-weather-model/

Oh, P. S., & Oh, S. J. (2011). What teachers of science need to know about models: An overview. *International Journal of Science Education*, *33*(8), 1109–1130. https://doi.org/10.1080/09500693.2010.502191

Park, B.-Y., Rodriguez, L., & Campbell, T. (2019). Using models to teach science. *The Science Teacher*, *87*(4), 8–11.

Park, J., Teo, T. W., Teo, A., Chang, J., Huang, J. S., & Koo, S. (2023). Integrating artificial intelligence into science lessons: Teachers' experiences and views. *International Journal of STEM Education*, *10*(61), 1–22. https://doi.org/10.1186/s40594-023-00454-3

Pedaste, M., Mäeots, M., Siiman, L. A., de Jong, T., van Riesen, S.A.N., Kamp, E. T., Manoli, C. C., Zacharia, Z. C., & Tsourlidaki, E. (2015). Phases of inquiry-based learning: Definitions and the inquiry cycle. *Educational Research Review*, *14*, 47–61. https://doi.org/10.1016/j.edurev.2015.02.003

Phan, T., Yukhymenko, M., & Bonk, C. J. (2025). *AI and critical thinking: Students' uses of Ask.SMILE for question-generating practices* [Manuscript submitted for publication]. Kremen School of Education and Human Development, California State University, Fresno.

Prince, M., & Felder, R. (2007). The many faces of inductive teaching and learning. *Journal of College Science Teaching*, *26*(5), 14–20.

Shakurnia, A., Aslami, M., & Bijanzadeh, M. (2018). The effect of question generation activity on students' learning and perception. *Journal of Advances in Medical Education & Professionalism*, *6*(2), 70–77.

Teig, N., Scherer, R., & Nilsen, T. (2018). More isn't always better: The curvilinear relationship between inquiry-based teaching and student achievement in science. *Learning and Instruction*, *56*, 20–29. https://doi.org/10.1016/j.learninstruc.2018.02.006

U.S. Food and Drug Administration. (2025). *Approved list of artificial intelligence-enabled medical devices*. https://www.fda.gov/medical-devices/software-medical-device-samd/artificial-intelligence-enabled-medical-devices

Weiner, S., Lake, R., & Rosner, J. (2024). *AI is evolving, but teacher prep is lagging: A first look at teacher preparation program responses to AI*. Center on Reinventing Public Education. https://crpe.org/wp-content/uploads/Teacher-Prep-AI-2024.pdf

Wenger, E. (1998). *Communities of practice: Learning, meaning, and identity*. Cambridge University Press.

Wenger-Trayner, E., & Wenger-Trayner, B. (2015). *An introduction to communities of practice: A brief overview of the concept and its uses*. Available from authors at https://www.wenger-trayner.com/wp-content/uploads/2022/06/15-06-Brief-introduction-to-communities-of-practice.pdf

Windschitl, M., Thompson, J., & Braaten, M. (2018). *Ambitious science teaching*. Harvard Education Press.

Wood, D., Bruner, J. S., & Ross, G. (1976). The role of tutoring in problem-solving. *Journal of Child Psychology and Psychiatry*, *17*, 89–100. https://doi.org/10.1111/j.1469-7610.1976.tb00381.x

Yen, T. S., & Halili, S. H. (2015). Effective teaching of higher order thinking (HOT) in education. *The Online Journal of Distance Education and e-Learning*, *3*, 41–47. https://tojdel.net/journals/tojdel/articles/v03i02/v03i02-04.pdf

Yu, F.-Y. (2009). Scaffolding student-generated questions: Design and development of a customizable online learning system. *Computers in Human Behavior*, *25*(5), 1129–1138. https://doi.org/10.1016/j.chb.2009.05.002

Zrada, M., Kastens, K. A., & Turrin, M. (2019). How to help your students ask more and better questions. *Oceanography*, *32*(4), 204–206. https://www.jstor.org/stable/26845653

9

ENHANCING STEM EDUCATION THROUGH AI-INFUSED SUSTAINABILITY LESSONS

The Design and Impact of an Asynchronous Online Professional Development for Middle and High School Science Teachers

Tiffany A. Roman and Brendan E. Callahan

Introduction

In 1989, the American Academy for the Advancement of Science (AAAS) highlighted several critical global issues, including unchecked population growth, environmental pollution, the shrinking of tropical rainforests, and social strife. These problems, compounded by the extreme inequities in wealth distribution and the significant resources devoted to warfare, presented a daunting challenge for humanity. Despite the passage of more than three decades, these issues remain pressing and, in many cases, have intensified, resulting in a situation where problems emerge faster than solutions can be found.

In 2011, the AAAS emphasized the crucial role of biologists in tackling these global challenges. The organization advocated the importance of science in society, the need to educate students about pressing global issues, and the preparation of a future workforce capable of addressing these challenges. That is, scientific literacy today requires an understanding of societal issues from multiple perspectives to facilitate effective reasoning and problem-solving (Kumar et al., 2024). This underscores the ongoing relevance of AAAS's earlier warnings and the need for a concerted effort to address these worsening global challenges (AAAS, 2011).

Incorporating sustainability concepts throughout the secondary science curriculum can help prepare students for citizenship in a changing world. Specifically, teaching sustainability can help prepare students to address pressing environmental, social, and economic challenges by developing their knowledge, attitudes, and behaviors across dimensions of sustainable development (Esa, 2010; Pauw et al., 2015). Sustainability education can also support other more general competencies and skills, such as critical thinking, problem-solving, and systems-level understanding (Weatherhead, 2020). In short, by teaching students to deal with interconnected complex systems and situations, sustainability education can equip them to better navigate a broad range of challenges (Zoller, 2012).

If sustainability education is going to be successful, active involvement of teachers is crucial (Agirreazkuenaga, 2019). However, many teachers believe they lack the knowledge,

DOI: 10.4324/9781003530343-10

skills, and confidence necessary to effectively integrate sustainability into their teaching (Green & Somerville, 2014). Yet others have simply not been trained to teach with sustainability concerns in mind (Groening, 2022). Therefore, it is important to provide teachers with appropriate training and support to enhance their capacity to teach sustainability (Anyolo et al., 2018; Fischer et al., 2022).

One way to address this need is through research-based continuous professional development (Casanova et al., 2023; Groening, 2022). Toward that goal, the authors provided online professional development to practicing science teachers over the course of two semesters geared toward incorporating sustainability concepts into the curriculum. To aid them in designing their instruction, the teachers were introduced and encouraged to explore how artificial intelligence (AI) tools could support the design of sustainability lessons. The following question guided the study:

> When supported by continuous online professional development, what types of sustainability practices do secondary science teachers emphasize in the science lessons they design?

Program Design Strategies for STEM Online Teacher Professional Development

To design meaningful STEM online teacher professional development focused on sustainability, the authors used adult learning and social constructivist theories (Guba & Lincoln, 1994). Effective teacher professional development is known to include the following characteristics: "(a) focus on content knowledge; (b) opportunities for active learning; and (c) coherence with other learning activities" (Garet et al., 2001, p. 916). It also encompasses structural features that further enhance teacher learning, such as the form of professional development (e.g., a series of workshops), duration of the professional development, and the collective participation of teachers who teach in the same content area, at the same grade level, and/or work within the same school (Garet et al., 2001). Where possible, professional development should be job-embedded and situated in the schools where participants work (Postholm, 2012; Powell & Bodur, 2019). It is also critical that teachers find the content useful and that it will inform their practice (Dede et al., 2009). Finally, effective professional development is attentive to educational reforms, context, curriculum, and collaboration (Sancar et al., 2021). Table 9.1 highlights these research-based professional development strategies, along with examples of how the strategies were incorporated into the online professional development.

Beyond these general characteristics of effective professional development, when hosted in an online modality, teachers value professional development that is relevant, authentic, useful, collaborative, interactive, reflective, and contextual (Powell & Bodur, 2019). Teachers, generally, are receptive to online professional development. For example, in a survey of 387 teachers who participated in a formal online professional development activity, Meyer et al. (2023) found that participants reported high levels of cognitive activation and moderate levels of collaboration. As a result, the authors recommended that when designing online teacher professional development, cognitive activation and collaboration must combine to create changes in teachers' professional practices.

TABLE 9.1 Summary Table of Online PD Design

Research-Based PD Strategies Used	*How We Integrated Strategy in the PD*
Integrate Adult Learning Theories	Adult learners are self-directed and value problem-centered, hands-on learning aligned to their work needs. To ensure the PD addressed teachers' needs, teachers completed a needs analysis survey, which helped us identify their existing sustainability knowledge and skills, as well as their experiences and challenges teaching sustainability. The survey informed the design of our learning modules. The teachers were encouraged to tailor their project-based deliverables to their science content areas, grade-level curriculum, and school context. Learners could work at a time that fits with their schedules asynchronously online.
Focus on Useful and Relevant Content Knowledge That Informs Practice	We included a sustainability expert as part of the professional development team for relevant and up-to-date content knowledge. We also connected content to Next Generation Science Standards (NGSS) and the UN Sustainable Development Goals (SDG). Through the integration of an AI module that encompassed awareness of AI, evaluating AI tools for classroom use, and using AI to support student-centered learning, we sought to provide relevant and timely technological content knowledge grounded in the teachers' instructional practice.
Opportunities for Active Learning	The teachers engaged in biweekly online asynchronous discussions and engaged in project-based learning focused on developing sustainability lessons to be used with their students. Additional active learning strategies involving peer communication transpired through the sharing of final projects where teachers had opportunities to provide their peers with asynchronous feedback.
Coherence Across Learning Activities	We utilized intentional design to structure professional development. The first PD module introduced the foundations of sustainability; the second explored the concepts of environmental literacy and systems thinking. These two modules provided the background for further study into critical aspects of sustainability: climate change and biodiversity, as well as energy and resource management. Module 5 connected AI learning to sustainability, as the teachers participated in the Friday Institute training (see below), and made connections to their own teaching. The final module provided an opportunity for teachers to share their sustainability lessons for feedback and refinement.
Structural PD Features (Form, Duration, Collective Participation)	The structure of the PD supported the busy summer schedules of adult learners; therefore, using an asynchronous, time-bound (6-week) PD format was intentional. To support a streamlined user experience, Microsoft (MS) Teams served as a "base camp" for participant communication and artifact sharing. MS Teams worked well for teachers who were not affiliated with the university. The university's learning management system (LMS) was not an option for hosting the PD as the LMS requires users to have net IDs and two-factor authentication, whereas Teams enabled easy access for users outside the institution. The learning modules were designed and hosted in Articulate 360. Articulate 360 is the industry standard software for online training as it has a professional interface and interactive features. One module focused on AI was hosted through NC State's Friday Institute. The collective participation in the training was supported through the participants' shared experiences. All middle and high school science teachers in the PD earned the same MAT degree from the same institution and all worked within the same metropolitan area, though at different schools.

(Continued)

TABLE 9.1 (Continued)

Research-Based PD Strategies Used	*How We Integrated Strategy in the PD*
Job-Embedded and Authentic Situated Learning	Deliverables included lesson and curriculum plans developed by teachers for use and implementation with their secondary science students in their fall courses. Many of the discussion prompts also focused on the participants' teaching practices and contexts.
Attention to Educational Reforms, Context, and Curriculum	NGSS and UN SDGs served as the context for learning objectives.
Cognitive Activation and Collaboration	Initial discussion prompts activated prior knowledge and revealed beliefs about sustainability. Further discussions encouraged participants to consider current and future classroom practices.
Reflection Opportunities	Discussion prompts focused on reflection. Prompts at the beginning of modules were designed to elicit prior knowledge; prompts at the end of modules were designed for participant reflection on their learning.

Learner-Centered Training and Structure of Sustainability Content

Based on recommended practices for teacher professional development, we created a learner-centered professional development designed to introduce teachers to sustainability concepts and the United Nations Sustainable Development Goals (SDGs) (United Nations, 2015). The training consisted of six modules that took place over two semesters. The course was offered in an asynchronous format by three faculty members: a professor of biology education, an associate professor of instructional technology, and an associate professor of biology. An undergraduate student assistant enrolled in the Learning Design and Teaching program at the authors' institution was hired to help manage the professional development platform, Articulate Rise 360, used to host the learning and reflection modules. The six modules included in the workshop focused on the following topics:

- Module 1: Foundations of Sustainability
- Module 2: Environmental Literacy and Systems Thinking
- Module 3: Climate Change and Biodiversity
- Module 4: Energy and Resource Management
- Module 5: AI in the Classroom [designed by the Friday Institute for Educational Innovation at North Carolina State University]
- Module 6: Communicating Sustainability in the Classroom [deliverables submitted through Microsoft Teams; teacher participants shared projects and completed a final discussion board post]

The first four modules consisted of a set of readings and videos to provide background knowledge of the topic and the underlying science. Each of these modules culminated in an online discussion board where the teachers responded to prompts related to the topic. The fifth module consisted of the teachers taking an AI workshop for teachers developed by The Friday Institute for Educational Innovation at North Carolina State University, which can be viewed at https://place.fi.ncsu.edu/local/catalog/course.php?id=46&ref=1

Within that module, the teachers watched four webinars, experimented with AI tools (ChatGPT, Canva's Magic Media), created artifacts that were both personal and practical, engaged in a series of reflection questions, set professional goals related to AI, and assessed new AI tools (e.g., Gamma, Parlay Genie, Curipod, Diffit, QuestionWell) using a suggested framework from the Friday Institute. Finally, the sixth module required the teachers to complete a deliverable for the professional development series, which focused on a lesson sequence related to sustainability (Table 9.2).

TABLE 9.2 Details from the Professional Development

Module	*Objectives*	*Resources and Description*
Module 1: Introduction/The Foundations of Sustainability	Describe the concept of sustainability. Explore the three pillars of sustainability: environmental, social, and economic. Discuss the importance of sustainable practices in the classroom.	We utilized some introductory readings, including the United Nations Sustainability Agenda and a STEM teaching tool, as well as a video from UCLA to support teachers' basic understanding of sustainability. This was to get everyone to use the same terminology and to have a similar background understanding of the term. **Pre-module discussion prompt:** What is your current understanding of sustainability? What have you done to include sustainability concepts in your science or mathematics teaching? **Post-module discussion prompt:** Now that you have learned a little bit about sustainability, share your experiences with sustainability in your daily life. What sustainable practices do you partake in either in your personal life or in your classroom? What are the things you currently do to try to teach sustainability in your classroom? Are there any topics that you have learned so far or will learn as a part of this workshop that interest you?
Module 2: Environmental Literacy and Systems Thinking	Develop an awareness of environmental literacy. Apply systems thinking to analyze complex sustainability issues. Investigate ecological interconnections with an emphasis on the human dimension.	This module introduced participants to the concept of environmental literacy, which relates to both the natural world and humans' interactions with the planet. Systems thinking is a methodology for examining complex issues by looking at the relationships and connections between systems of interest. Systems and system models are one of the crosscutting concepts in the NGSS, yet it is usually taught superficially. We used a variety of videos and readings, ranging from practitioner-based to research articles, on utilization of systems thinking in K-12 settings to generate support for utilizing systems thinking in their own classrooms. **Deliverable:** Consider a science topic that could incorporate sustainability ideas and share with the group.

(Continued)

TABLE 9.2 (Continued)

Module	*Objectives*	*Resources and Description*
Module 3: Climate Change and Biodiversity	Examine the scientific evidence for climate change. Read and classify biodiversity and its role in ecosystems. Develop strategies for teaching climate-related topics.	Climate change is an important concept, as there are many misconceptions surrounding the topic. For example, many people conflate weather and climate and erroneously wonder why the weather is cold if the planet is warming. With this module, we wanted to present scientific evidence for climate change, as well as link the impact of climate change on biodiversity. We included videos from TEDEd and the BBC to introduce biodiversity and climate change and then incorporated the STEM Teaching Tools (https://stemteachingtools.org/) to provide ideas for incorporating these concepts in the classroom. STEM Teaching Tools are short briefs on a variety of teaching topics that can be used to provide ideas for teaching concepts consistent with NGSS. **Post-module discussion:** For this activity, select TWO of the climate learning resources from the STEM Teaching Tools webpage and report on them. Give a very short (2–3 sentence) summary of the tool, as well as ideas for how you can incorporate the ideas from the teaching tool into your practice for the upcoming year.
Module 4: Energy and Resource Management	Differentiate the variety of renewable energy sources. Describe what efficient resource utilization looks like (water, energy, materials). Select energy and resource concepts for inclusion into science lessons.	While many sustainability efforts are concentrated on what individuals can do, we wanted participants to gain an awareness of the larger system that supplies the world with the energy it needs. In addition to considering the environmental impacts of energy sources, we examined the economic and social aspects of resource management. We used videos from the BBC and a TEDTalk by a technologist to discuss clean energy infrastructures and resource management. A short article that discusses UN SDG Goal 7: Affordable and Clean Energy also provides some insight into the importance of energy supplies. **Post-module discussion:** How do the points made in the TEDTalk correspond to the current narrative about renewable resources? How can we encourage our students to advocate for more renewable resource availability?
Module 5: AI in the Classroom	Learn about AI's potential in education. Begin the journey to AI fluency. Set a goal for continued learning around AI in education.	AI is transforming education in a variety of ways; however, many teachers are not equipped to use AI themselves or prepare their students to utilize AI. Participants took NC State's Friday Institute for Educational Innovation's "AI with the FI" online PD.

(Continued)

TABLE 9.2 (Continued)

Module	*Objectives*	*Resources and Description*
		Their stated goal is "to empower educators, offering hands-on guidance to foster AI fluency and transform teaching and learning." The course objectives included: • Learn about AI's potential in education • Develop your AI fluency as an educator • Explore responsible AI use • Find efficiency and purpose for using AI in your practice • Identify key strategies for responsible AI in your classroom • Align AI with learner needs, academic integrity, and content standards • Learn from the AI experience and expertise of other educators • Advocate for AI as a digital leader **Deliverables:** Submit your completed workbook and AI-infused lesson plan.
Module 6: Communicating Sustainability in the Classroom		This module was designed for participants to reflect on their experiences in the program as well as share their finalized sustainability-infused lesson plans.

Methods

Research Methodology

To examine the phenomenon within this study, which encompassed the sustainability practices emphasized by secondary science teachers when supported by continuous online teacher professional development, the authors employed a descriptive qualitative case study method (Stake, 1995). Within a case study, one examines the phenomenon within a bounded system in action (Miles et al., 2020), with factors related to location, sampling, or time. In the context of the present study, the bounded context was a cohort of six middle and high school science teachers who had previously participated in a Master of Arts in Teaching (MAT) program specific to secondary science education and an English as a second language endorsement from a regional public research university in the southeastern United States. The unit of analysis was the phenomenon of sustainability education as experienced by the teachers in a 100% online continuous professional learning environment.

To better support the trustworthiness of the study, we prioritized the qualitative research criteria set forth by Shenton (2004) and Guba (1981). The criteria included credibility (thorough literature review, multiple data sources, thick description, member checking), transferability (thick description), dependability (overlapping data collection methods, audit trail, two researchers coding the data), and confirmability (data triangulation, multiple methods, collecting data from multiple sources).

Study Context

The study was conducted during a two-semester-long professional development (summer and fall 2024) designed for practicing secondary science teachers. All participants had earned a Bachelor's degree in a science content area and a MAT degree. The professional development was designed to be relevant for the participants based on a needs assessment completed prior to the development of the workshop. The results of the needs assessment indicated that the teachers had a solid understanding of the three dimensions (Disciplinary Core Ideas, Science and Engineering Practices, and the Crosscutting Concepts) of the Next Generation Science Standards (NGSS) (NGSS Lead States, 2013) but were less knowledgeable about the United Nations SDGs. When asked specifically about using AI, some of the teachers reported having used AI applications infrequently but being interested in learning more about the potential for AI integration for ethical student use.

> At my school, AI has not really been introduced or incorporated. I have personally sought out resources with AI to generate starter questions, real-life applications, and review games for students.
>
> *(Emma)*

> As far as students using AI, I would love to hear about how others may be using AI with their students as a tool instead of a way to cheat. It would be great to see if anyone has any ideas about teaching students to use it responsibly.
>
> *(Chloe)*

Overall, the teachers in the cohort were familiar with AI, had been introduced to it in educational contexts, but did not have enough knowledge to effectively incorporate it into their teaching practices. They also were not comfortable with incorporating AI for student use, which should ultimately be the goal of AI-infused instruction.

The authors sought to provide flexibility to the teachers by offering opportunities to engage with their peers in multiple ways through synchronous opportunities and asynchronous communication. Due to the participants' busy summer and fall schedules, most interactions were completed asynchronously. Asynchronous discussion boards provided teachers with the opportunity to reflect on their readings before responding to their peers.

Participants

The participants were six members of a professional development experience designed for practicing secondary science teachers who had previously graduated from the participating university. All were Robert Noyce stipend recipients. The Robert Noyce Teacher Scholarship Program offers financial support to higher education institutions to provide scholarships, stipends, and other resources aimed at attracting and training STEM majors and professionals to become K-12 teachers. The goal of the program is to boost the number of K-12 teachers working in high-need school districts with robust STEM expertise. The participants were exceptional undergraduate students in biology who had completed a MAT degree as part of the requirements for their initial teaching licensure. At the time of the study, they were practicing teachers with between three and eight years of experience. They represented both middle grades (two) and high school (four) from five different Title 1 schools in the metro-Atlanta area.

Data Collection

The authors utilized discussion boards during the professional development to examine teachers' ideas about sustainability in general as well as their integration of sustainability concepts into their teaching practice. Additionally, to examine their evolving perspectives and priorities when planning instruction, data were collected on the teachers' lesson sequence on sustainability as well as documentation from their AI workshop. In total, there were 39 of a possible 42 original posts for the discussion boards, with additional responses to original posters. The discussion boards were copied and pasted from the course shell by the first author. The posts were then uploaded to ATLAS.ti to enable analysis of the teachers' perspectives on sustainability and teaching with AI.

Data Analysis

The authors uploaded the discussion boards as well as the lesson sequences produced by the participants to perform qualitative analyses for each of the constructs under investigation. Within each post and document, we examined references to sustainability, such as the United Nations SDGs, elements of sustainability, and so on. Additionally, we examined how the teachers were utilizing AI in their teaching practices. The second author investigated the documents for elements of sustainability, whereas the first author initially investigated the documents for references to AI. The two authors then discussed their coding schemes to reach consensus on the themes.

Findings

Prioritizing Sustainability in the Classroom

Prior to the workshop, the teachers demonstrated some variability in their definitions of sustainability but tended to prioritize the need to consider both the future and the present when making decisions, whether for future generations or to maximize long-term benefits. They also included in their definitions the idea of resource management and of minimizing harm to the environment and society. Below are three examples of participants' definitions of sustainability, as gathered from the workshop:

> Sustainability means to live in a way that supports the needs of our world today while also ensuring our ability to meet our needs in the future.
>
> *(Alice)*

> Sustainability, to me, means considering the long-term impacts of our actions and looking ahead. It involves conserving our resources while making space for growth and technological advancements that can lead to a better, more sustainable future.
>
> *(Daniel)*

The following views were reflected in the teachers' discussion of the sustainability concepts they included in their teaching lessons prior to participating in professional development. While some comments reflected more student-centered work, others were more teacher-centered. There was also a particular focus on reducing human impact on the environment. Select participant quotes are provided below.

> We do a human impacts project where students examine impacts of humans on the environment and propose ways to shrink that impact or reverse damage.
>
> *(Chloe)*

> We focus on sustainability at the beginning of the year with our ecology unit and analyzing human impact on ecological processes.
>
> *(Fiona)*

> We plan to teach sustainability through hands-on learning, particularly through gardening and various projects. As the sponsors of the garden club, my colleague and I oversee our greenhouse, and we plan to involve students in practical activities related to planting and reusing natural resources.
>
> *(Daniel)*

While some of the teachers mentioned including sustainability concepts in their standard teaching practices, some of their responses indicated that they also incorporated special projects in the curriculum to emphasize sustainability concepts, such as renewable energies and community involvement. In addition to teaching lessons that explicitly incorporated sustainability concepts, the teachers also included sustainable practices in their teaching. One of the common sustainable practices in classrooms and schools was recycling. Teachers were also utilizing more online or paperless activities to reduce the amount of waste being produced. Below, examples of these findings are presented.

> In my school we have a team of students who head up recycling. We recycle in my classroom and at home.
>
> *(Chloe)*

> My school has both white paper and cardboard recycling that I utilize, but I would like to implement plastic bottle recycling as well. I try to limit the amount of paper I use in class by having my openers, warmups, closers, ticket-out-the-doors, and knowledge check quizzes on our LMS [learning management systems] and then recycling the paper I do hand out.
>
> *(Bella)*

While recycling is a common household and classroom practice that encourages sustainability, some teachers incorporated other practices to widen their students' exposure to sustainable practices based on their contexts or experiences. Examples of expanded sustainability practices included the following:

> I encourage students to practice energy conversation and to use items found in and around the house to complete class activities.
>
> *(Alice)*

> During labs, I am particular as to what should be reused, and I teach them [students] the concept that many labs they may be a member of in the future will have limited resources and part of their responsibility is to conserve those consumables and resources as much as possible to save their company money.
>
> *(Bella)*

The first quote is from a teacher who was teaching virtual classes, so it makes sense that her classroom is paperless. The second quote comes from a teacher who had made a career change after working in a laboratory setting, so she was particularly aware of cost-saving measures taken by businesses that also support sustainability goals.

Planning for a Sustainable Future

As the main deliverable for the workshop, the teachers were instructed to develop a lesson sequence that included sustainability concepts they could use in their own class context. Table 9.3 summarizes these lessons under the name of each participant.

As illustrated, the lessons aligned well with both the NGSS and the United Nations SDGs; however, most of them centered on a limited variety of Science and Engineering Practices (SEP), Crosscutting Concepts (CCC), and SDGs. Over half of the six lesson plans included the SEP of "Asking Questions and Defining Problems;" two included the SEPs of "Developing and Using Models" and "Obtaining, Evaluating, and Communicating Information." A similar trend was found for the CCCs, as three of the lessons included the CCC of "Cause and Effect" and "Systems and System Models;" "Energy and Matter" was the only other CCC that was included in more than one lesson plan.

Regarding the inclusion of an AI module, collectively, the teachers who completed the activities developed by the Friday Institute found the resources to be worthwhile, as illustrated in the following examples.

> The AI with the FI [Friday Institute] workshop was incredibly valuable in introducing me to a wide range of AI tools. I'm excited to explore these further and incorporate them into future lesson planning.
>
> *(Alice)*

> I really enjoyed learning about how to move towards advocacy with AI. There were so many resources introduced; along with varying ways to implement AI within the classroom that I look forward to trying in the future.
>
> *(Emma)*

More specifically, within the AI workbooks that they completed, participants detailed which new AI tools they had assessed, as well as their future professional learning goals related to AI. One teacher, Bella, who was already familiar with the generative AI capabilities of ChatGPT and Canva's Magic Media opted to explore a Chrome extension called Brisk to "see about helping increase the differentiation in my lesson plans for my ESOL students." Other teachers reported using tools like CuriPod, Parlay Genie, Diffit, and Socratic to support lesson planning and student learning.

The AI workbooks provided insights into how generative AI factored into the professional development deliverables the teachers created for the sustainability workshop. In

TABLE 9.3 Alignment of Secondary Science Teacher Participant Lessons with the Next Generation Science Standards and United Nations Sustainable Development Goals

Topic and Class	*NGSS – DCI*	*NGSS – SEP*	*NGSS – CCC*	*UNSDGs*
Alice "Sustainable Seas: Navigating Human Impact" HS Biology	HS-LS2-7: Design, evaluate, and refine a solution for reducing the impacts of human activities on the environment and biodiversity.	Asking questions and defining problems Obtaining, evaluating, and communicating information		7 – Affordable and clean energy 12 – Responsible consumption and production 13 – Climate action 14 – Life below water
Bella "Sustainable Biofuels: Innovating for a Greener Future" HS Biotechnology	HS-PS3-3: Design, build, and refine a device that converts one form of energy into another. HS-LS2-7: Design, evaluate, and refine a solution for reducing the impacts of human activities on the environment and biodiversity. HS-ETS1-3: Evaluate a solution to a complex real-world problem based on prioritized criteria and trade-offs.	Asking questions and defining problems Developing and using models Planning and carrying out investigations Constructing explanations and designing solutions	Stability and change Scale, proportion, and quantity Cause and effect Systems and system models	7 – Affordable and clean energy 12 – Responsible consumption and production 13 – Climate action
Chloe "Hot Compost" HS Biology	HS-LS1-6: Construct and revise an explanation based on evidence for how carbon, hydrogen, and oxygen from sugar molecules may combine with other elements to form amino acids and/or other large carbon-based molecules. HS-LS1-7: Use a model to illustrate that cellular respiration is a chemical process whereby the bonds of food molecules and oxygen molecules are broken and the bonds in new compounds are formed, resulting in a net transfer of energy.	Asking questions and defining problems Developing and using models	Energy and matter Cause and effect Systems and system models	2 – Zero hunger 12 – Responsible consumption and production

Daniel "Sustainability and Greenhouse Design" HS Biology	HS-ESS3-3: Create a computational simulation to illustrate the relationships among the management of natural resources, the sustainability of human populations, and biodiversity. HS-LS2-7: Design, evaluate, and refine a solution for reducing the impacts of human activities on the environment and biodiversity.			7 – Affordable and clean energy 12 – Responsible consumption and production 13 – Climate action
Emma "Driving Towards Cleaner Skies" MS Physical Science	MS-ESS3-3: Apply scientific principles to design a method for monitoring and minimizing human impact on the environment. MS-PS1-2: Analyze and interpret data on properties of substances before and after the substances interact to determine if a chemical reaction has occurred. MS-PS1-6: Undertake a design project to construct, test, and modify a device that either releases or absorbs thermal energy by chemical processes.		Cause and effect Systems and system models Energy and matter	3 – Good health and well-being 7 – Affordable and clean energy 9 – Industry, innovation, and infrastructure
Fiona "Becoming Agents for Change through the Sustainable Development Goals" MS Life Science	Variable	Asking questions and defining problems Obtaining, evaluating, and communicating information	Variable	Variable – all SDGs introduced

some cases, generative AI provided teachers with ideas on how to expand their learning activities with their students. For example, one teacher wrote:

> I'm about to start a waves unit with my students, so I asked ChatGPT to create a mini lesson for me about waves suited for 8th-grade students. It presented the lesson plan (unprompted) with the 5E model, with activities incorporated in each. I did enjoy the layout of this; it gave me ideas of interactive ways I can introduce waves that I had not previously used. It really sets a nice baseline for me or anyone that just needs to see something structured out first or that has what they want but need a little more direction to see it come to fruition.
>
> *(Emma)*

In this example, the use of ChatGPT broadened the teacher's perspective on potential learner activities that she had not considered before. By comparison, another teacher, Alice, was more critical of the output of the AI tool that she had used (Curipod), noting that a lesson generated for her with learner activities on the topic of meiosis was "very basic" and lacked "any mention of the individual phases." However, she found it "not terribly problematic because I have my own slides with the correct information that I can add into the lesson." This same participant used AI technology with her students during the implementation phase of her sustainability lesson to support formative feedback for her learners. Reflecting on the experience, she wrote:

> I have used a Curipod lesson with my students since beginning Module 1. It worked mostly like a Nearpod from the student point of view. However, after students answered a short-answer question, the Curipod gave them AI feedback on the response. I polled the students after each lesson, and they gave it an average of 4/5 stars. Most said they thought the AI feedback was cool. While I always give them feedback on their answers during lessons using instructional technology like Nearpod, it was nice to have the AI give the students individualized feedback. Every AI response I saw was detailed and accurate. It was also very encouraging in its constructive criticism.
>
> *(Alice)*

A common theme was how participants' perceptions of generative AI shifted after watching an Educator AI Showcase webinar that provided authentic examples from educators detailing how AI was used in their classrooms to support student learning. One teacher, Bella, wrote that she now considered, "how AI can be integrated into the curriculum in various ways where the students can utilize it for problem-solving, research, and critical thinking not just completing the assignment for them." Likewise, another teacher, Bella, wrote, "I like that they [Educator AI Showcase] showed varying ways/tools to implement in the classroom that empower student learning and encourage collaboration with AI instead of relying heavily on a tool to do the work for students." Others spoke about a shift in their own advocacy for integrating AI for student-centered use, with one participant noting:

> I plan to lead by example. I want to integrate AI tools in ways that directly benefit my students while staying transparent about the process. My next step is to create a project

where students use AI tools to solve real-world problems, helping them see AI as a resource they can trust and critique. I'll also share my experiences with colleagues, encouraging them to explore and adopt AI in their own classrooms.

(Daniel)

In sum, there was a notable shift from perceiving generative AI as a mechanism that outputs work to viewing it as a collaborative learning agent that can support student learning and collaboration. In at least one instance, a participant directly mentioned leading by example and sharing their experiences with colleagues at their school.

Conclusions and Recommendations

This chapter detailed the design and implementation of an asynchronous in-service secondary science teacher professional development enacted over two semesters. Based on our findings, we argue that online professional development efforts in STEM teacher education require a sustained collaborative team effort if the aim is to provide meaningful, innovative, and authentic learning experiences. Aspects of successful professional development include flexibility, relevance, and rapport among participants as well as with the facilitators (Dille & Røkenes, 2021).

Importance of Teacher Rapport, Flexibility, and Content-Area Relevance for the Success of Professional Development Training

This asynchronous online STEM professional development experience benefited from working with teachers who had previously established rapport with one another. The authors had provided other professional development opportunities to the six teachers in the past, and some of the teacher participants were from the same cohort of MAT graduate students. Furthermore, the teacher participants also had a rapport with the science education faculty member, who taught in the MAT program and had worked with them previously. The existing rapport between the faculty and a subset of teachers, as well as participants' familiarity with one another, is notable.

A key premise of this professional development was built on previously established trust with the teachers to support their ongoing learning situated in the classroom. That is, in addition to content and pedagogical knowledge, professional development facilitators should enact emotional and social *embodiment,* a term that encompasses a positive and collaborative rapport with learners in professional development settings (Perry & Booth, 2024). In this instance, with a positive rapport previously established, we were able to create time and space for the participants to become reacquainted with one other and to introduce new faculty at the onset of the professional development. These steps were beneficial in creating an environment where the participants would feel successful.

Regarding flexibility and relevance, a critical component of the success of this online STEM professional development stemmed from its intentional design to reflect teachers' needs and availability to engage in the training. That is, we took a value-based design approach (Korthagen, 2017) to addressing teachers' needs by carrying out a needs assessment prior to the development of the modules. For example, the teachers articulated a lack of knowledge related to AI and the United Nations Sustainability Development Goals. As a result, the training focused on these topics, which resulted in relevant, innovative lessons developed and implemented by the teachers in fall 2024. Additionally, given that professional development started during the summer, we sought to prioritize flexibility, designing professional learning to be available on demand and reflective of the needs of adult learners. As a result, all learning modules were fully asynchronous to better support the needs of the teachers who opted into the training, which was not required by their respective schools or the districts in which they worked.

Recommendations

Recommendations for online STEM professional development designers based on this study are threefold. First, although job-embedded professional development is considered ideal (Powell & Bodur, 2019), it is possible to offer meaningful online STEM professional development to teachers that share an existing interest, connection, and/or background experience. The participants in this professional development opportunity were all Noyce Scholars and shared the experience of being trained as STEM majors and professionals prior to becoming middle and high school teachers in different Title I schools across the metro-Atlanta area. Though they could not have a job-embedded professional development experience together, they were open to engaging in professional development on their own time in the summer and fall. Further, they were more than willing to actively participate in discussions, share ideas, and actively engage throughout the program. As recommended by Casanova et al. (2023), we provided teachers with agency in their learning and supported the co-construction of knowledge with their peers. For teacher educators interested in providing a similar STEM professional development, an asynchronous online format that enables teacher agency and peer feedback can capitalize on existing rapport and shared background experience teachers already have, allowing facilitators to hit the ground running.

A second recommendation for online STEM professional development: It is important to design the professional development around teachers' needs by introducing relevant and practical content knowledge designed by a multidisciplinary team. As noted, the faculty who participated in designing and developing this online training came from the fields of instructional technology, biology, and biology education. To aid in the production of professional learning activities, an undergraduate Learning, Design, and Technology student built out the learning modules in Articulate 360 based on the content and pedagogy recommended by the faculty members. The collective knowledge and skills brought by the team was critical to the success of the project, as it would not have been possible without the input of all team members. Furthermore, by intentionally creating a space for collaboration and sharing in Microsoft Teams, teacher participants were able to reach out to the faculty directly, engage in asynchronous discussions, and share their sustainability lessons with one another.

As a result of the design of the professional development, as well as the introduction of meaningful and relevant content knowledge, the authors observed shifts in science teachers' self-efficacy regarding teaching sustainability and integrating AI into student learning. Teacher participants not only created lessons that aligned with the NGSS and the United Nations SDGs, but they also engaged in a research-based AI professional learning module developed by North Carolina State University's Friday Institute and were able to design curricula that incorporated AI teaching and learning.

This highlights the third recommendation based on this work: It is important to integrate existing high-quality STEM professional development resources when possible and appropriate. Gathering online STEM teacher professional development resources, developing learning modules, and facilitating learner activities is time consuming, particularly in online settings. As such, it is critical to strive to work smarter, not harder, to achieve the desired outcomes of any online STEM professional development.

There are known benefits of online teacher professional development, including changes to attitudes, beliefs, and self-efficacy, along with an increase in content knowledge (Morina et al., 2025). Online teacher professional development is further enhanced when teachers can use the skills and materials learned in the training in student-centered learning classroom contexts (Morina et al., 2025). This study exemplifies how online STEM professional development can be successful when (a) existing in-service teacher networks are engaged; (b) faculty expertise is varied and valued; and (c) existing research-based resources are integrated. Furthermore, by providing STEM teachers with flexibility in their schedules, not only will teachers voluntarily engage in meaningful and relevant STEM professional development opportunities, but they may also lead by example in their school communities, modeling to their colleagues what they learned in their online STEM professional development.

KEY TAKEAWAYS

- Create online STEM professional development for teachers who share an existing interest, connection, and/or background experience.
- Design online professional development based on STEM teachers' needs, providing teachers with agency to co-construct knowledge.
- Introduce relevant and practical content knowledge designed by a multidisciplinary team.
- Consider integrating existing, high-quality STEM professional development resources, where possible.

References

Agirreazkuenaga, L. (2019). Embedding sustainable development goals in education. Teachers' perspective about education for sustainability in the Basque autonomous community. *Sustainability, 11*(5), 1496. https://doi.org/10.3390/su11051496

American Academy for the Advancement of Science (AAAS). (1989). *Science for all Americans: Education for a changing future*. Oxford University Press.

American Academy for the Advancement of Science (AAAS). (2011). *Vision and change in undergraduate biology education: A call to action*.

Anyolo, E. O., Kärkkäinen, S., & Keionen, T. (2018). Implementing education for sustainable development in Namibia: School teachers' perceptions and teaching practices. *Journal of Teacher Education for Sustainability, 20*(1), 64–81. https://doi.org/10.2478/jtes-2018-0004

Casanova, C. R., King, J. A., & Fischer, D. (2023). Exploring the role of intentions and expectations in continuing professional development in sustainability education. *Teaching and Teacher Education, 128*, 104115. https://doi.org/10.1016/j.tate.2023.104115

Dede, C., Jass Ketelhut, D., Whitehouse, P., Breit, L., & McCloskey, E. M. (2009). A research agenda for online teacher professional development. *Journal of Teacher Education, 60*(1), 8–19. https://doi.org/10.1177/0022487108327554

Dille, K. B., & Røkenes, F. M. (2021). Teachers' professional development in formal online communities: A scoping review. *Teaching and Teacher Education, 105*, 1–17. https://doi.org/10.1016/j.tate.2021.103431

Esa, N. (2010). Environmental knowledge, attitude and practices of student teachers. *International Research in Geographical and Environmental Education, 19*(1), 39–50. https://doi.org/10.1080/10382040903545534

Fischer, D., King, J., Rieckmann, M., Barth, M., Büssing, A., Hemmer, I., & Lindau-Bank, D. (2022). Teacher education for sustainable development: A review of an emerging research field. *Journal of Teacher Education*, *73*(5), 509–524.

Garet, M. S., Porter, A. C., Desimone, L., Birman, B. F., & Yoon, K. S. (2001). What makes professional development effective? Results from a national sample of teachers. *American Educational Research Journal*, *38*(4), 915–945. https://doi.org/10.3102/00028312038004915

Green, M., & Somerville, M. (2014). Sustainability education: Researching practice in primary schools. *Environmental Education Research*, *21*(6), 832–845. https://doi.org/10.1080/13504622.2014.923382

Groening, Z. (2022). *The effect of an intervention on experienced chemistry teachers' pedagogical content knowledge of environmental sustainability* (Publication No. 30700041) [Doctoral Dissertation, University of Pretoria]. ProQuest Dissertations & Theses Global.

Guba, E. G. (1981). Criteria for assessing the trustworthiness of naturalistic inquiries. *Educational Communication and Technology*, *29*(2), 75–91. https://www.jstor.org/stable/30219811

Guba, E. G., & Lincoln, Y. (1994). Competing paradigms in qualitative research. In N. Denzin & Y. Lincoln (Eds.), *Handbook of qualitative research* (pp. 105–117). Sage.

Korthagen, F. (2017). Inconvenient truths about teacher learning: Towards professional development 3.0. *Teachers and Teaching*, *23*(4), 387–405. https://doi.org/10.1080/13540602.2016.1211523

Kumar, V., Choudhary, S. K., & Singh, R. (2024). Environmental socio-scientific issues as contexts in developing scientific literacy in science education: A systematic literature review. *Social Sciences & Humanities Open*, *9*, 100765.

Meyer, A., Kleinknecht, M., & Richter, D. (2023). What makes online professional development effective? The effect of quality characteristics on teachers' satisfaction and changes in their professional practices. *Computers & Education*, *200*, 104805. https://doi.org/10.1016/j.compedu.2023.104805

Miles, M. B., Huberman, A. M., & Saldana, J. (2020). *Qualitative data analysis: A methods sourcebook* (4th ed). Sage.

Morina, F., Fütterer, T., Hübner, N., Zitzmann, S., & Fischer, C. (2025). Effects of online teacher professional development on teacher, classroom, and student level outcomes: A meta-analysis. *Computers & Education*, 105247. https://doi.org/10.1016/j.tate.2018.09.004

NGSS Lead States. (2013). *Next Generation Science Standards: For states, by states*. https://www.nextgenscience.org/

Pauw, J. B.-d., Gericke, N., Olsson, D., & Berglund, T. (2015). The effectiveness of education for sustainable development. *Sustainability*, *7*(11), 15693–15717. https://doi.org/10.3390/su71115693

Perry, E., & Booth, J. (2024). The practices of professional development facilitators. *Professional Development in Education*, *50*(1), 144–156. https://doi.org/10.1080/19415257.2021.1973073

Postholm, M. B. (2012). Teachers' professional development: A theoretical review. *Educational Research*, *54*(4), 405–429. https://doi.org/10.1080/00131881.2012.734725

Powell, C. G., & Bodur, Y. (2019). Teachers' perceptions of an online professional development experience: Implications for a design and implementation framework. *Teaching and Teacher Education*, *77*, 19–30. https://doi.org/10.1016/j.tate.2018.09.004

Sancar, R., Atal, D., & Deryakulu, D. (2021). A new framework for teachers' professional development. *Teaching and Teacher Education*, *101*, 103305. https://doi.org/10.1016/j.tate.2021.103305

Shenton, A. K. (2004). Strategies for ensuring trustworthiness in qualitative research projects. *Education for Information*, *22*(2), 63–75. https://doi.org/10.3233/EFI-2004-22201

Stake, R. E. (1995). *The art of case study research*. Sage.

United Nations. (2015). *Transforming our world: The 2030 agenda for sustainable development*. https://sdgs.un.org/sites/default/files/publications/21252030%20Agenda%20for%20Sustainable%20Development%20web.pdf

Weatherhead, N. K. (2020). *Integrated techniques to increase systems thinking for sustainability education of high school science students using educational design research* (Publication No. 28155073). [Doctoral Dissertation, University of South Carolina]. ProQuest Dissertations & Theses Global].

Zoller, U. (2012). Science education for global sustainability: What is necessary for teaching, learning, and assessment strategies? *Journal of Chemical Education*, *89*(3), 297–300. http://doi.org/10.1021/ed300047v

10

INTEGRATING HAZARD-RESILIENT HOUSING INTO STEM CLASSROOMS

Using the Model of Authentic Learning in Online Professional Development

Thomas Chandler, Joshua De Vincenzo, and Shuyang Huang

Introduction

As wildfires, hurricanes, floods, and extreme heat waves become more pronounced due to climate change, the built environment in which U.S. communities reside is increasingly threatened (National Academies of Sciences, Engineering, and Medicine, 2021). Residential housing infrastructure, in particular, is under enormous strain, with only one-third of American homes currently adhering to the latest building codes (Strupp, 2023). Entire regions face new threats due to poor structural integrity of buildings for flooding, inadequate fire suppression systems, and a lack of retrofitting to withstand hurricanes or high wind events. Such limited attention to housing mitigation underscores an urgent need for a new generation of learners to understand both the challenges and the opportunities related to hazard-resilient design (American Planning Association, 2019). Such designs take a holistic approach that impacts all phases of a building's lifecycle, with an emphasis on building performance and minimizing the potential negative impacts of the building on occupants and the environment (U.S. General Services Administration, 2025).

Despite the enormous economic and societal cost associated with increasing risks to the U.S. housing stock, K-12 subject matter has only nominally addressed housing sustainability within science, technology, engineering, and mathematics (STEM) school curricula. That is, while educators may teach about extreme weather threats conceptually, these concepts are taught within science subject matter, rather than being extended into authentic and observable design-based solutions for sustainable living. Although housing is a real-world theme for many educators, as most have firsthand experience with the comforts and constraints of the place they call "home," contemporary pedagogy and textbooks do not place emphasis on integrating it into STEM subject matter. This results in the loss of classroom discussions about housing safety, affordability, and equity in building construction, along with the growing risks to the U.S. housing supply due to the increase in natural disasters (Holmlund et al., 2018).

Given that most educators have not had the opportunity to learn about hazard-resilient housing or how to incorporate the topic into their classroom instruction, providing

DOI: 10.4324/9781003530343-11

professional development (PD) to this audience is essential (Will & Prothero, 2023) to ensure future progress in this area. This chapter examines how online PD can be designed to help K-12 educators integrate housing sustainability themes into their STEM learning environments. Authentic Learning is introduced as the guiding conceptual framework for designing such online PD, drawing on its late-20th-century educational foundation and its relevance in fields such as architecture and urban planning, where professionals frequently collaborate virtually. The chapter goes on to outline an application framework that includes a table of online learning tools and their instructional uses. It concludes with suggestions for instructional designers and PD providers on how to prepare educators to incorporate housing issues brought on by climate change via technologically enhanced learning opportunities.

Instructional Design Framework: Model of Authentic Learning

Situated learning theories from the late 20th century, especially those of Lave and Wenger (1991) and Brown et al. (1989), which all emphasize that knowledge is best comprehended in the context of its application, are the foundation of Authentic Learning. To provide a more comprehensive definition, Herrington et al. (2010) identified nine key elements, including cooperation, reflection, authentic activities, and authentic settings. According to their model, students become more involved and are able to apply their knowledge and skills to professional practice when they work on tasks that mimic real-world issues. However, for Authentic Learning to take place, the delivery of instruction must also be authentic.

Authentic Learning is a compelling model for PD because it views educators as active producers of their own content rather than passive recipients. Instead of focusing on abstract training modules, Authentic Learning-based PD programs encourage educators to engage in demanding, real-world projects that enhance their teaching approaches and procedures. For instance, educators might be requested to collaborate on developing STEM simulations, multimedia resources, or digital lesson plans that reflect the complexity of their own classrooms (Herrington & Parker, 2013). These design-based activities have the potential not only to improve educators' technological and pedagogical skills but also to assist them in producing practical resources that they can immediately implement in their classrooms.

Web-based technologies expand the possibility for actual learning in PD by offering spaces for both production and collaboration. Using virtual tools, educators can develop interactive exercises and prototype teaching resources, or show students how to solve problems that are relevant to their learning. This is in line with the constructionist perspective (Papert, 1993), which maintains that producing artifacts with personal meaning is the most effective approach to producing knowledge. By situating design tasks in real-world learning environments, PD programs can encourage higher levels of cooperation, engagement, and transferability, which, in turn, provides educators with more opportunities to implement the knowledge they acquire in PD into their digital teaching methods (Rule, 2006). Finally, educators have grown more accustomed to online and hybrid learning environments since the COVID-19 pandemic, thus increasing their receptivity to digital methods. New online resources allow educators to visualize the scale of climate risks, manipulate design variables, simulate disaster scenarios, and help their students collaborate in real time.

The model of Authentic Learning is built around nine key elements (Herrington, 2006; Herrington & Oliver, 2000) that help ensure learning tasks mirror the complexity, collaboration, and reflection needed in real-world problem-solving. They are listed as follows and then discussed in detail below.

1 Provide authentic contexts that reflect the way the knowledge will be used in real life.
2 Provide authentic activities.
3 Provide access to expert performances and the modeling of processes.
4 Provide multiple roles and perspectives.
5 Support collaborative construction of knowledge.
6 Promote reflection to enable abstractions to be formed.
7 Promote articulation to enable tacit knowledge to be made explicit.
8 Provide coaching and scaffolding by the teacher at critical times.
9 Provide for an authentic assessment of learning within the tasks.

Application of the Model of Authentic Learning to Housing Sustainability Instruction in Online Professional Development

Authentic online K12 PD environments place educators in tasks that mimic the intricacies of real-world practice. To help them ground their work in issues that are similar to those encountered by architects and urban planners, a workshop on hazard-resilient housing design, for example, could introduce them to geospatial data and local hazard maps (Herrington et al., 2010). Participants understand how their work directly relates to classroom application and community resilience when learning is embedded in real-world contexts. By ensuring that PD is not abstract but connected to urgent local and global issues such as hazard mitigation and climate change, this contextual grounding boosts educators' confidence in their ability to teach these subjects.

Another aspect for fostering authenticity involves engaging participants in authentic tasks, such as extended projects that require sustained effort and critical thinking. Educators might be tasked with collaboratively designing a digital prototype of a hazard-resilient home or community space using accessible design platforms such as SketchUp, QGIS, or ArcGIS StoryMaps. Educators can then work with students on understanding system-level connections between housing and factors such as socioeconomic effects, building codes, community histories, and societal roles in disaster risk reduction, all with the aim of identifying solutions. These projects embody the type of complex, ill-structured problems that characterize Authentic Learning (Rule, 2006). By integrating content knowledge with design and problem-solving skills, educators experience inquiry-based processes that they can later implement with their own students.

Authentic online PD workshops also emphasize exposure to expert performance and multiple perspectives. Guest presentations by architects, urban planners, or climate scientists provide models of professional practice and situate the workshop within broader disciplinary expertise. Meanwhile, group activities allow participants to assume different stakeholder roles, such as engineers, policymakers, or community members, reflecting the multi-perspective collaboration that characterizes real-world hazard-mitigation projects (Herrington & Parker, 2013). This dual exposure to experts and diverse viewpoints supports educators in understanding both the technical and social dimensions of sustainable design. For instance, educators could be introduced to basic engineering topics such as load distribution, material strength, and thermal efficiency, which directly affect the resilience and sustainability of housing. Using online simulation tools such as Energy3D, PhET simulations, or introductory Building Information Modeling (BIM) software, educators can experiment with variables such as insulation materials, structural geometry, or energy transfer. Further, through scaffolded tutorials and collaborative digital workspaces, participants can model hazard-resistant structures, such as homes elevated against flooding or those designed with renewable energy integration, ensuring that engineering is not taught abstractly but through meaningful, practice-based projects.

Collaboration, presentations, reflection, and scaffolding function as interrelated elements that guide participants through the design process. Collaborative digital workspaces and breakout groups foster collaboration, while opportunities for presentations, including discussing design rationales, encourage teachers to make their abstract reasoning more concrete. Reflection, facilitated through guided prompts or online journaling, helps participants critically evaluate their design choices and their instructional implications. Scaffolding by facilitators, whether through tutorials, case studies, or just-in-time feedback, ensures that educators remain supported without undermining the authenticity of the task (Herrington et al., 2010). Together, these processes create an environment where professional growth is both collaborative and deeply individualized.

Lastly, authentic assessment anchors the PD experience by evaluating the participants' deliverables against professional standards. Rather than relying on traditional tests, assessment focuses on the design prototypes themselves, judged by criteria such as hazard resilience, creativity, and educational utility. Peer review and self-assessment provide additional layers of evaluation, encouraging educators to adopt reflective and critical stances toward their work. This approach ensures that assessment is integrated into the learning process, reinforcing the principle that authentic PD should not only enlighten but also transform instructional practices (Herrington et al., 2010).

Table 10.1 lists the nine elements of the Authentic Learning model, along with complementary examples of activities, teaching strategies, and tools for fostering virtual collaboration about hazard-resilient housing design themes.

TABLE 10.1 Elements of the Authentic Learning Model Applied to Online PD Activities, Strategies, and Tools

Nine Elements	*Activities*	*Suggested Strategies to Teach Online*	*Tools*
1. Authentic contexts	Explore real hazard maps and local housing vulnerabilities	Use case studies from participants' regions Integrate local data. Overlay climate hazard layers (floodplains, wildfire risks, sea-level rise projections) on local housing data	ArcGIS Online, QGIS, Google Earth Engine, FEMA Flood Map Service
2. Authentic activities	Design a hazard-resilient house or community layout using digital tools	Assign project-based challenges with real-world constraints. Explore professional-level housing design workflows, including sustainability plug-ins	SketchUp, Tinkercad, Revit (intro level)

(*Continued*)

TABLE 10.1 (Continued)

Nine Elements	*Activities*	*Suggested Strategies to Teach Online*	*Tools*
3. Expert performance	Watch professionals model hazard-mitigation design strategies	Invite guest speakers or show recorded design walkthroughs. Model housing designs and test daylighting, shading, ventilation, or energy efficiency PD developers can introduce real-world workflows used by architects and urban planners, reinforcing Authentic Learning's emphasis on professional practice	Video conferencing program (e.g., Zoom, Microsoft Teams), YouTube case study videos
4. Multiple roles and perspectives	Analyze housing resilience from engineering, environmental, and cross-cultural comparison	Facilitate breakout rooms where each group adopts a stakeholder perspective	Miro, Padlet, Google Docs
5. Collaboration	Work in interdisciplinary teams to co-design digital prototypes	Use group projects with shared online workspaces	Google Workspace, Video conferencing program (e.g., Zoom, Microsoft Teams), Slack, Trello
6. Reflection	Write or record reflections on design choices and challenges	Use reflective journals or discussion forums with guiding prompts	Moodle learning management system (e.g., Google Classroom), Padlet
7. Articulation	Present design prototypes and rationale to peers	Require team presentations and recorded video pitches	Video conferencing program (e.g., Zoom, Microsoft Teams), Microsoft PowerPoint or Google Slides
8. Coaching and Scaffolding	Receive facilitator feedback and resources during the design process	Provide just-in-time tutorials, checklists, and sample designs	Short screencasts, Curated readings
9. Authentic assessment	Evaluate the final digital design for hazard resilience and pedagogical relevance	Use rubrics aligned to real-world criteria Peer- and self-assessment components	Rubric tools in Canvas, Google Forms for peer feedback

Content-Oriented Online Tools for Housing Sustainability in STEM

To translate the principles of Authentic Learning into practice, PD providers and instructional designers can also integrate a variety of content-oriented online learning tools. These tools enable visualization, simulation, collaboration, and iterative design, all of which are critical components of effective STEM instruction on housing sustainability. While not directly addressing all nine elements of the Authentic Learning model, the tools in Table 10.2 particularly support three principles: (a) providing authentic contexts, (b) providing authentic activities, and (c) providing expert performance.

TABLE 10.2 Online Learning Tools for Housing Sustainability in STEM

Tool	*Description*	*Application in Housing Sustainability Education*
Mapping and Geospatial Analysis		
ArcGIS Online: https://www.arcgis.com	Cloud-based mapping and analytics platform that enables users to create, share, and analyze interactive maps and spatial data.	PD providers can use ArcGIS StoryMaps to combine virtual local hazard maps with narrative explanations, helping participants visualize how climate risks affect housing sustainability. Students can then create their own story-driven virtual projects that propose resilient housing solutions tailored to their communities.
ArcGIS StoryMaps: https://storymaps.arcgis.com	Web-based platform that combines interactive maps with narrative text, images, and multimedia to create engaging, story-driven presentations.	PD participants can create and share interactive online maps that combine hazard data, demographic information, and land-use patterns. These online maps can then form the basis for participant-designed housing solutions that are socially and environmentally responsive.
Google Earth Engine: https://earthengine.google.com	Cloud-based platform for planetary-scale geospatial analysis using satellite imagery and environmental data.	This virtual software tool allows PD participants to work with large-scale satellite imagery and climate datasets to view and share long-term environmental changes affecting housing sustainability. PD providers can guide participants in using these datasets to propose design adaptations for resilient communities.
QGIS: https://qgis.org	Free, open-source geographic information system (GIS) that allows users to visualize, analyze, and edit geospatial data.	This software can be used by PD providers to teach participants how to map and analyze environmental and hazard data, such as flood zones or wildfire risks, in relation to housing sites. Participants can then design and share virtual housing plans that respond to these mapped vulnerabilities.
FEMA Flood Map Service https://msc.fema.gov/portal/home	Provides U.S. government flood hazard maps and data used to assess flood risk and support emergency management and planning.	This U.S. government online service provides official flood risk maps that can be integrated into PD workshops on hazard-resilient housing. PD participants can explore their own communities' flood exposure and design housing solutions that mitigate these risks.
Simulation and Energy Analysis		
Energy3D: https://energy.concord.org	Simulation-based software that allows users to design, model, and analyze the energy performance of buildings and renewable energy systems.	This software enables PD participants to model the energy efficiency of buildings by experimenting with orientation, insulation, and renewable energy features. PD providers can assign live virtual design challenges in which learners test how sustainable housing designs perform under different environmental conditions.

(Continued)

TABLE 10.2 (Continued)

Tool	*Description*	*Application in Housing Sustainability Education*
3D Design and Modeling		
Autodesk Revit BIM (Building Information Modeling) https://www.autodesk.com	Software program used by architects and engineers to design, document, and simulate building structures and systems in 3D.	Introduces PD participants to professional-level Building Information Modeling (BIM), where they can collaboratively design and test building systems for sustainability and hazard resilience. PD providers can use simplified projects to demonstrate online how energy efficiency and structural stability are integrated into housing design.
Rhino (Rhinoceros 3D) + Grasshopper Plugin: https://www.rhino3d.com	3D modeling software for design and architecture, and its Grasshopper plugin enables parametric modeling and algorithmic design through visual programming.	This software enables advanced parametric modeling in which PD participants can virtually simulate structural responses to environmental factors like wind, shading, or energy use. In online instruction, PD providers can assign live virtual design challenges where participants test different sustainable housing configurations digitally.
SketchUp for Schools: https://edu.sketchup.com/app	Web-based 3D modeling program designed for K-12 use, enabling students to design and share models directly within Google Workspace or Microsoft Education accounts.	With this software, PD participants can design 3D models of homes that integrate features like passive solar orientation, rainwater harvesting, or green roofs. Online lessons can include peer review sessions in which participants share and refine their designs collaboratively, online.
Tinkercad: https://www.tinkercad.com	Free, browser-based 3D design and coding platform that allows users to create simple models, circuits, and simulations for prototyping and STEM learning.	This software can be used for introductory 3D modeling tasks in which PD participants prototype simple sustainable housing elements, such as solar panel layouts or wind-resistant roof structures. PD providers can scaffold projects to connect these basic prototypes to larger online housing design challenges.
Game-Based and Interactive Learning		
PhET Interactive Simulations: https://phet.colorado.edu	Free, research-based online tools that allow students to explore and visualize science and math concepts through interactive experiments.	PD providers can incorporate PhET simulations to virtually demonstrate scientific principles such as heat transfer, solar energy, and structural forces. These concepts can then be applied in online design projects where participants evaluate housing features that improve disaster resilience and energy efficiency.
Minecraft Education Edition: https://apps.apple.com/us/app/minecraft-education/id1196524622	Classroom-friendly version of the game that supports project-based learning by enabling students to build and explore virtual worlds with STEM and sustainability concepts.	With this software, PD participants can design 3D models of homes that integrate features like passive solar orientation, rainwater harvesting, or green roofs. Online lessons can include peer review sessions in which participants share and refine their designs collaboratively, online.

Conclusion

Increases in the frequency and intensity of natural disasters pose a grave threat to the U.S. housing stock. As communities confront these challenges, it is imperative that educators, in particular those teaching STEM subjects, are equipped with the competency and capability to imagine and teach sustainable, resilient solutions. Yet, issues of hazard-resilient housing remain largely absent from K-12 STEM education.

Online PD offers an opportunity to bridge this gap. Through Authentic Learning, rooted in late-20th-century digital instruction approaches, educators can engage students in authentic, collaborative problem-solving that mirrors the practices of architects, engineers, and urban planners. By leveraging online tools, such as 3D modeling platforms and mapping applications, PD participants can design housing prototypes, test their ideas, and present solutions that are both creative and grounded in scientific understanding. Further, these concepts can then be implemented with their own students.

For K-12 PD developers and instructional designers, incorporating housing sustainability into STEM curricula is not only feasible but also transformative. It empowers participants to see themselves as designers and problem solvers, capable of addressing global challenges through innovation and collaboration (Blessing & Ross, 1996). As educators embrace this opportunity, they will prepare their students not only to excel in STEM disciplines but also to contribute to more effective designs for the built environment.

KEY TAKEAWAYS

- U.S. housing infrastructure is increasingly vulnerable to climate-related hazards such as floods, wildfires, and hurricanes, yet housing sustainability concepts are largely absent from K-12 STEM curricula. Educators need support to bring these critical, real-world issues into their classrooms. PD opportunities offer an innovative and effective way to do this.
- This chapter positions the model of Authentic Learning as an effective framework for online PD. Its nine elements, such as authentic context, collaboration, reflection, and authentic assessment, all help ensure that educators engage in tasks that mirror real-world professional practice.
- Rather than treating educators as passive learners, the PD model described in this chapter situates them as active designers of digital artifacts. Activities include creating prototypes of hazard-resilient housing, developing multimedia resources, and designing lesson plans that embed engineering and sustainability concepts.
- A wide range of digital platforms, from 3D modeling software such as SketchUp and Tinkercad to geospatial tools such as ArcGIS StoryMaps, can be used to support authentic, design-based PD. These tools enable collaboration, visualization of climate risks, and simulation of building performance under different hazard scenarios.

Acknowledgment

The authors wish to thank the National Center for Disaster Preparedness, Columbia University student intern Chloe Chung for conducting background research in relation to this chapter.

References

American Planning Association. (2019). *Planning for resilience: Addressing climate risks in housing and communities*.

Blessing, S. B., & Ross, B. H. (1996). Content effects in problem categorization and problem solving. *Journal of Experimental Psychology: Learning, Memory, and Cognition*, 22(3), 792.

Brown, J. S., Collins, A., & Duguid, P. (1989). Situated cognition and the culture of learning. *Educational Researcher*, *18*(1), 32–42. https://doi.org/10.3102/0013189X018001032

Herrington, J. (2006). Authentic e-learning in higher education: Design principles for Authentic Learning environments and tasks. In T. C. Reeves & S. Yamashita (Eds.), Proceedings of E-Learn 2006 (pp. 3164–3173). AACE.

Herrington, J., & Oliver, R. (2000). An instructional design framework for Authentic Learning environments. *Educational Technology Research and Development*, *48*(3), 23–48. https://doi.org/10.1007/BF02319856

Herrington, J., & Parker, J. (2013). Emerging technologies as cognitive tools for Authentic Learning. *British Journal of Educational Technology*, *44*(4), 607–615. https://doi.org/10.1111/bjet.12048

Herrington, J., Reeves, T. C., & Oliver, R. (2010). *A guide to authentic e-learning*. Routledge.

Holmlund, T., Lesseig, K., & Slavit, D. (2018). Making sense of 'STEM education' in K-12 contexts. *International Journal of STEM Education*, *5*(1), 32.

Lave, J., & Wenger, E. (1991). *Situated learning: Legitimate peripheral participation*. Cambridge University Press.

National Academies of Sciences, Engineering, and Medicine. (2021). *Accelerating decarbonization of the U.S. energy system*. National Academies Press. https://doi.org/10.17226/25932

Papert, S. (1993). *The children's machine: Rethinking school in the age of the computer*. Basic Books.

Rule, A. C. (2006). The components of Authentic Learning. *Journal of Authentic Learning*, *3*(1), 1–10.

Strupp, J. (2023, September 14). Building codes aren't climate ready, but changes are coming. *Construction Dive*. https://www.constructiondive.com/news/building-codes-climate-resilience-nibs/693672

U.S. General Services Administration. (2025). *Sustainable design*. https://www.gsa.gov/real-estate/design-and-construction/sustainability/sustainable-design#

Will, M., & Prothero, A. (2023, September 18). U.S. teachers lag behind global peers in teaching about sustainability. *Education Week*. https://www.edweek.org/teaching-learning/u-s-teachers-lag-behind-global-peers-in-teaching-about-sustainability-heres-why/2023/09

11

EXTENDED TPACK FOR ONLINE PROFESSIONAL DEVELOPMENT

Addressing the Dual-Layer Challenge of STEM Teacher Educators

Woonhee Sung

Introduction

Online professional development (PD) has emerged as a critical component of contemporary teacher education, offering more accessible and sustainable opportunities for learning and skill building. As educational systems increasingly recognize the potential of digital platforms to deliver high-quality, sustained professional learning experiences, the design and implementation of online PD programs have become central to efforts aimed at improving teaching practices and student outcomes. Science, technology, engineering, and mathematics (STEM) education, in particular, requires educators to understand and implement integrated approaches that connect STEM content in meaningful ways. The complexity and interdisciplinary nature of STEM make it difficult for PD providers and instructors to design interactive online PDs. Adding to this challenge, most K-12 teachers currently engaged in supporting STEM learning activities were trained within their own subject discipline when pursuing their teacher education degrees, making them potentially less fluent in implementing the integrative and cross-disciplinary approaches advocated in STEM education (Jong et al., 2021; Lo, 2021).

Studies of STEM-focused PD programs have demonstrated that participants in online formats rate their experiences as equally valuable and effective as face-to-face iterations and highlighted the benefits of integrating technology components in online formats over face-to-face delivery (Fies & Packham, 2021). Thus, when designed effectively, online delivery offers unique affordances that can enhance PD outcomes, particularly in developing teachers' technological competencies for their own classroom practice.

This is why it is important that online PD targeting STEM teachers promote an integrative STEM teaching approach with effective uses of technologies and pedagogies that can be transferred into teachers' classroom teaching. That is, when PD instructors deliver online programs, they must demonstrate effective online learning facilitation and simultaneously prepare teachers for diverse K-12 classroom contexts extending beyond traditional "teacher-to-student" relationships to "teacher educator-to-teacher-to-student" dynamics. This means that there is another layer on top of the teacher-student relationship when we

DOI: 10.4324/9781003530343-12

design PD for teachers in teacher education settings. This complexity is particularly acute in STEM education, where teachers must integrate Interdisciplinary Content Knowledge (ICK) across science, technology, engineering, and mathematics while adapting pedagogical strategies and technological tools for classrooms and students.

This chapter addresses this "dual-layer" challenge faced by online PD designers/providers/instructors when teaching STEM content for K-12 teachers by introducing an extension of the foundational Technological Pedagogical Content Knowledge (TPACK) framework (Mishra & Koehler, 2006). While the TPACK framework provides solid guidance for equipping teachers with content as well as pedagogical and technological knowledge for their classroom teaching, the author has added a dual-layer approach to specifically guide online PD instructors. That is, building on Borko's (2004) recognition that PD is a complex system requiring careful attention to facilitator knowledge and context, this chapter proposes extensions to existing TPACK components to address dual-layer complexities through (a) *ICK*, which encompasses authentic connections between STEM domains; (b) *Extended Pedagogical Knowledge (EPK)*, which distinguishes between facilitation strategies for adult learners in online contexts and implementation strategies for K-12 students; and (c) *Extended Technology Knowledge (ETK)*, which supports both interactive online PD delivery and diverse classroom technology integration needs.

In addition, the extended TPACK (E-TPACK) framework ensures accessibility, scalability, and meaningful integration of interdisciplinary content with pedagogical and technological expertise. Drawing on established PD principles (Darling-Hammond et al., 2017; Guskey, 2000) and TPACK components discussed in Mishra and Koehler (2006), we present the E-TPACK framework with application examples that leverage the unique affordances of online delivery while addressing its challenges.

TPACK as Conceptual Framework

In addressing the integration of technology, pedagogy, and content knowledge, The TPACK framework (Mishra & Koehler, 2006) identifies three major knowledge domains – Content Knowledge (CK), Pedagogical Knowledge (PK), and Technology Knowledge (TK) – along with their intersections, which lead to the core, where all three domains overlap (Koehler & Mishra, 2009).

Petko et al. (2025) recently highlighted the influence of external factors, so-called "Contexts," that surround the core components, leading to the recognition that TPACK is situated and must be understood within specific educational contexts (Petko et al., 2025). Moreover, a recent study has revealed a distinction between knowledge that develops within specific learning environments and contextual knowledge about the diverse contexts where that learning will be implemented (Petko et al., 2025). This distinction becomes particularly critical in online PD, where facilitators must master online teaching, engagement, and technology implementation strategies while simultaneously preparing teachers for varied K-12 classroom implementations.

The TPACK framework's strength lies in its holistic perspective, recognizing that effective technology integration requires not only technical skills but also an understanding of the relationship between technological tools, pedagogical approaches, and CK (Schmidt et al., 2009; Voogt et al., 2013). TPACK has demonstrated practical applicability across diverse educational contexts, providing a concrete structure for understanding and developing technology integration competencies valuable for both teacher education and PD (Chai et al., 2013; Rosenberg & Koehler, 2015). However, while the framework is backed by extensive research validating its utility and implementation guidance (Phillips, 2017; Wang et al., 2018), it does not directly support the context of PD, especially online PD, which poses a unique set of requirements.

When developing online PD that integrates STEM disciplines, PD providers/instructors need to consider (a) content domains, the knowledge across science, technology, engineering, and mathematics; (b) PK for online teaching and preparing K-12 teachers' teaching strategies; and (c) technological knowledge for online facilitation demonstrating diverse tools and for K-12 teachers' class usage. To address these three aspects, we propose the E-TPACK framework. This extended approach aims to provide a scalable solution for building STEM integration competency among teacher learners through meaningful PD experiences that bridge the gap between online learning environments and diverse K-12 classroom implementation contexts.

Extended TPACK

The E-TPACK framework directly responds to the three specific aspects required for effective online STEM PD: (a) the single-domain content assumption, (b) the classroom-specific pedagogical focus, and (c) the scope of TK. The framework operationalizes the dual-layer approach through three reconceptualized knowledge domains, each explicitly subdivided to address both facilitation and implementation contexts (see Figure 11.1).

Interdisciplinary Content Knowledge (ICK)

ICK extends the single-domain CK approach with an integrated understanding of authentic connections across science, technology, engineering, and mathematics. Rather than treating STEM disciplines as separate subjects to be coordinated, ICK emphasizes emergent understanding that arises when disciplinary boundaries dissolve through real-world problem-solving contexts (English, 2016; Kelley & Knowles, 2016). This approach enables PD providers to model authentic STEM integration while simultaneously preparing teachers to facilitate similar interdisciplinary experiences with their students.

Extended Pedagogical Knowledge (EPK)

EPK operationalizes the dual-layer pedagogical challenge through two distinct but interconnected knowledge domains: (a) Pedagogical Knowledge for Facilitation (PK-F) and (b) Pedagogical Knowledge for Application (PK-A). PK-F encompasses strategies for engaging adult learners in virtual STEM integration experiences. This includes creating online environments where interdisciplinary projects are delivered through interactive and manipulative formats accompanied by collaborative online problem-solving activities. Demonstration of diverse pedagogical strategies utilized in online PD sessions scaffolds teachers' understanding of how virtual activities can support authentic STEM integration in their own classrooms.

PK-A focuses on preparing teachers with PK applicable in diverse classroom contexts, not limited to online facilitation but more related to in-person K-12 settings. This includes integrative learning design principles, strategies for connecting students' prior knowledge

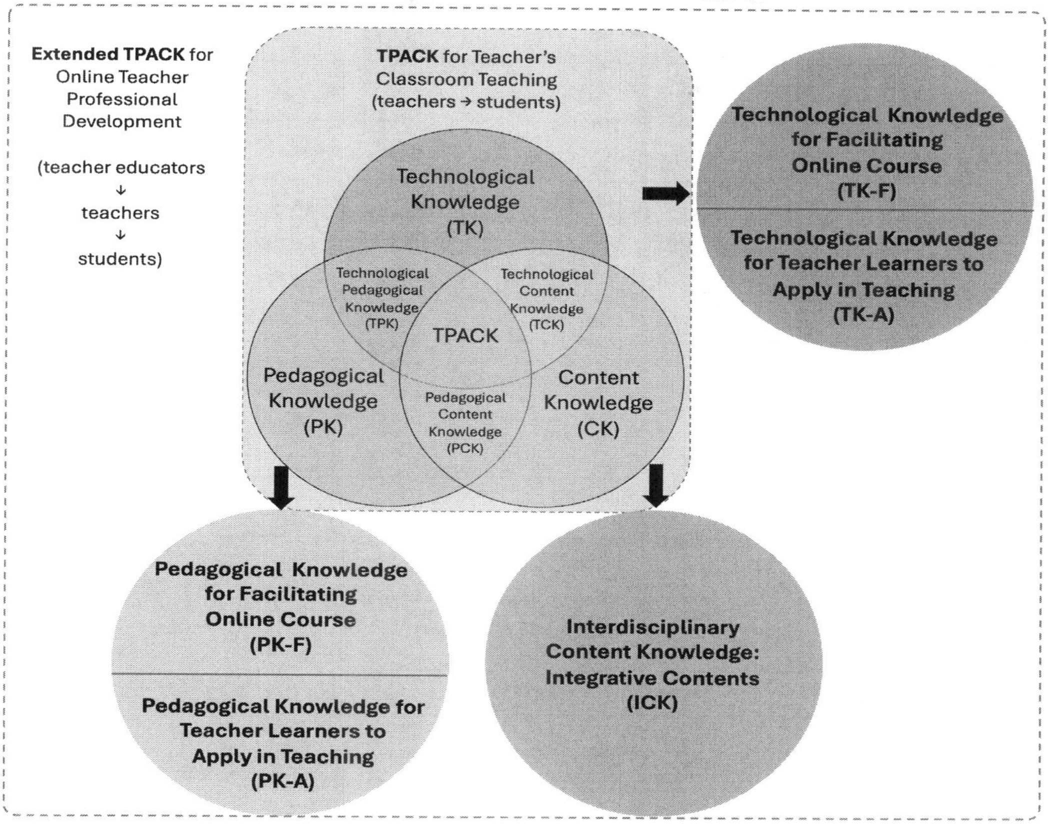

FIGURE 11.1 Extended TPACK (E-TPACK) Framework for Online STEM PD

across STEM domains, and approaches for managing the complexity of interdisciplinary instruction, as discussed in traditional K-12 STEM education. To ensure that teachers can adapt sophisticated integration approaches to their specific contexts, PK-A must account for varied classroom constraints, from resource limitations to diverse student populations.

Extended Technology Knowledge (ETK)

ETK addresses the technological complexity of operating across online facilitation and diverse implementation environments through (a) Technological Knowledge for Facilitation (TK-F) and (b) Technological Knowledge for Application (TK-A). TK-F includes expertise in digital platforms that enable authentic STEM project delivery in online contexts. This encompasses interactive simulation environments, collaborative design tools, virtual laboratory experiences, and online collaborative reflection/discussion tools. TK-F leverages the unique affordances of online delivery to provide rich technological experiences that would be impossible in traditional face-to-face formats.

TK-A focuses on classroom-appropriate technologies that maintain effectiveness across diverse resource environments. This includes software, applications, or online website resources that K-12 classrooms can offer and that can be used for STEM learning. Teachers need to learn technology-enhanced approaches for different resource levels, and strategies for maintaining pedagogical integrity.

Integration Areas and Central E-TPACK

The extended framework maintains TPACK's overlapping areas between domains, with each area addressing the dual-layer complexity of online PD. Table 11.1 lists E-TPACK's overlapping areas along with their descriptions and application in dual-layer contexts.

TABLE 11.1 E-TPACK's Overlapping Domain Descriptions and Application in Dual-Layer Contexts

Original Domains	*E-TPACK's Overlapping Domains*	*Description*	*Dual-Layer Application*
Pedagogy + Content Knowledge (PCK)	Extended Pedagogical Content Knowledge (E-PCK)	Combines interdisciplinary content understanding with dual-layer pedagogical expertise	Enables online PD providers to integrate pedagogies for facilitation with interdisciplinary STEM content knowledge while preparing teachers to develop their own pedagogical content knowledge for authentic STEM integration in diverse K-12 classroom contexts
Technology + Content Knowledge (TCK)	Extended Technological Content Knowledge (E-TCK)	Enhances STEM content understanding through technology-mediated integration	Allows online PD providers to demonstrate technology-enhanced interdisciplinary STEM content delivery using both online facilitation tools and classroom-appropriate technologies while preparing teachers to select and integrate appropriate technological tools that support authentic STEM learning across varied resource environments
Technology + Pedagogy Knowledge (TPK)	Extended Technological Pedagogical Knowledge (E-TPK)	Optimizes technology use across facilitation and implementation contexts	Empowers online PD providers to orchestrate sophisticated technology-mediated pedagogical approaches for adult learning in virtual environments while modeling and preparing teachers to implement technology-enhanced pedagogical strategies that maintain effectiveness across diverse K-12 classroom contexts and resource levels
Central TPACK	Central E-TPACK	Mastery of complex, multi-component interdisciplinary STEM educational systems across all domains and contexts	Requires online PD providers to masterfully coordinate all extended knowledge domains – integrating facilitation pedagogies with application pedagogies, online technologies with classroom technologies, and interdisciplinary content knowledge – while demonstrating for teachers how to develop comprehensive STEM integration expertise that translates effectively into improved K-12 classroom practice

The central E-TPACK integration represents the most complex aspect of the framework, as PD providers must navigate the tension between providing rich, technology-enhanced learning experiences and preparing STEM teachers for K-12 classroom contexts. The framework provides concrete guidance for PD design that considers both the unique online contexts and the practical realities of diverse K-12 implementation environments, ensuring that the benefits of virtual delivery enhance rather than complicate classroom implementation.

Application Example 1: COVID-19 Global Spread Analysis Project

This section presents a concrete application of the E-TPACK framework through an online PD activity that demonstrates how the dual-layer approach can be systematically implemented. The COVID-19 Global Spread Analysis Project, conducted online with teacher participants, illustrates how online PD projects can simultaneously prepare teachers for diverse classroom implementations.

The project guided teachers through an authentic interdisciplinary investigation while explicitly preparing them to adapt similar approaches in their own classrooms. Beginning with the content foundation, teachers engaged with scientific literature about viral transmission, comparing historical and contemporary pandemic patterns to establish biological and geographic understanding. This foundation phase modeled how to introduce complex scientific concepts while considering diverse learner needs. Teachers then conducted geographic investigations using digital mapping tools, with each participant assigned specific world regions to explore. They collected real-world epidemiological data while navigating virtual environments, learning both content about regional variations in disease spread and techniques for guiding student research across different technological contexts.

During data analysis phases, teachers transformed collected information into statistical visualizations, practicing mathematical and scientific interpretation skills while considering grade-appropriate adaptations. The digital storytelling component challenged teachers to synthesize research into multimedia presentations that communicated complex scientific concepts through engaging narratives, demonstrating both advanced digital creation skills and strategies for supporting student science communication. Peer sharing enabled teachers to publish work on collaborative platforms and engage in professional review processes, experiencing sophisticated online collaboration tools while planning simpler classroom alternatives. The project concluded with implementation planning, where teachers reflected on their learning experience and designed adaptations appropriate for their specific classroom contexts, taking into consideration resource constraints, student populations, and curricular requirements.

This progression ensured that teachers experienced authentic interdisciplinary learning while explicitly developing the capacity to facilitate similar integrated experiences with their own students, embodying the dual-layer complexity central to E-TPACK. The systematic application of the framework components is illustrated in Table 11.2.

This systematic application demonstrates how the E-TPACK principles guided the selection of content, pedagogies, and technologies that serve dual purposes: providing interactive PD experiences while preparing teachers to implement similar approaches with varied resources. The project exemplifies how sophisticated online delivery can enhance rather than complicate classroom implementation when designed with explicit attention to

TABLE 11.2 E-TPACK Application Example: COVID-19 Global Spread Analysis Project

Project Phases	*Integrated Contents (ICK)*	*Pedagogies for Facilitation (PK-F)*	*Pedagogies for Application (PK-A)*	*Tools for Facilitation (TK-F)*	*Tools for Application (TK-A)*
Content Foundation	Biology (virus transmission), History (1918 pandemic), Geography (global patterns)	Guided reading with adult learners, Online discussion facilitation, Collaborative text analysis	Plan age-appropriate virus education, Scaffold complex scientific concepts, Connect historical and current events	• History.com articles • WHO Situation Reports • Online conferencing software for discussion	• Classroom science articles • Comparison charts • Student-friendly timelines
Geographic Investigation	Geography (regional differences), Biology (environmental factors), Social Studies (cultural impacts)	Virtual field trips, Structured data collection, Regional expertise sharing	Guide student virtual explorations, Teach geographic research skills, Facilitate student data collection	• Google Earth (https://earth.google.com/ • WHO COVID-19 dashboard • International databases	• Google Earth (Google Map) • Local health websites • Age-appropriate mapping tools
Data Analysis	Statistics (trend analysis), Mathematics (graphing), Science (epidemiological interpretation)	Collaborative spreadsheet work, Adult technical skill development, Peer review of analysis	Teach student data literacy, Scaffold statistical thinking, Guide graph interpretation	• Google Sheets (Graphing, charting) • Advanced charting functions using other tools • Statistical analysis tools	• Basic graphing tools • Student data entry templates • Simple chart creation apps
Digital Storytelling	Technology (multimedia creation), Language Arts (narrative writing), Science Communication	Model digital storytelling, Facilitate creative expression, Support technical learning	Teach multimedia skills, Guide narrative development, Support science communication	• Google Earth projects • Advanced multimedia integration • Professional presentation tools	• Basic presentation platforms • Storytelling creation media or hands-on drawing ideas
Peer Sharing	Communication (peer review), Technology (collaboration), Critical Thinking (evaluation)	Facilitate professional community, Model constructive feedback, Guide reflection	Teach peer review skills, Facilitate collaboration, Guide constructive criticism	• Padlet (https://padlet.com/) • Real-time collaboration on synchronous conference meetings (breakout rooms)	• Sharing platform (Google drive) • Basic feedback tools • Classroom collaboration apps
Implementation Planning	Pedagogy (adaptation strategies), Assessment (evaluation), Curriculum Design (standards alignment)	Support teacher reflection, Guide adaptation planning, Facilitate professional growth	Plan authentic assessment, Adapt for diverse learners, Align with standards	• Professional planning platforms • Standards databases • Assessment design tools	• Planning templates • Simple rubrics • Student portfolio tools

dual-layer complexity. Each phase integrated multiple STEM disciplines through authentic problem-solving while modeling pedagogical approaches that teachers could observe, experience, and adapt for their own diverse contexts.

Application Example 2: AI Literacy Through Geometry Classification

The second application of the E-TPACK framework involves an online PD activity that integrates artificial intelligence (AI) literacy with mathematics teaching. The AI Literacy Through Geometry Classification Project (Sung & Gunpinar, 2025) demonstrates how the dual-layer approach can make complex computational concepts accessible while preparing teachers for diverse classroom implementations without requiring sophisticated technology infrastructure.

The project guided teachers through authentic machine learning experiences using familiar geometric content while explicitly preparing them to implement similar unplugged AI activities in their own classrooms. Beginning with algorithmic thinking foundations, teachers engaged with a Code.org sorting game that introduced AI training concepts through interactive ocean trash classification. This foundation phase modeled how to introduce algorithmic decision-making while considering diverse technological access levels across K-12 settings. Teachers then explored geometric-shape classification by identifying similar features between randomly selected target shapes and others, learning both mathematical classification concepts and techniques for guiding student pattern recognition across different resource contexts.

During the explanation phase, teachers analyzed the connection between AI feature selection and mathematical categorization, practicing interdisciplinary content integration while considering grade-appropriate adaptations. The elaboration component challenged teachers to create explainable feature matrices that organize geometric properties for algorithmic classification, demonstrating both advanced computational thinking skills and strategies for supporting student logical reasoning. The evaluation phase involved AI role-playing activities where teachers tested their matrices by identifying new shapes following their created algorithms, experiencing authentic machine learning processes while planning simpler classroom alternatives. The project concluded with implementation reflection, where teachers considered how to adapt the unplugged approach for their specific classroom contexts, addressing resource constraints, student populations, and curricular alignment requirements.

This progression ensured that teachers experienced sophisticated computational thinking while explicitly developing the capacity to facilitate similar integrated experiences with their own students, embodying the dual-layer complexity central to E-TPACK while maintaining accessibility across diverse educational settings. The systematic application of framework components is illustrated in Table 11.3.

This systematic application shows how E-TPACK is applied in designing mathematics teacher PD for promoting AI literacy. The online PD taught AI concepts for teachers' CK that can be combined in mathematics and other subject matters with problem-solving approaches and used different tools. Although the project was delivered online, the entire approach can be implemented in in-person contexts with hands-on materials, demonstrating the framework's flexibility across delivery modalities while maintaining the dual-layer pedagogical focus essential to effective teacher PD.

TABLE 11.3 E-TPACK Application Example: AI Literacy Through Geometry Classification

Project Phases	*Integrated Contents (ICK)*	*Pedagogies for Facilitation (PK-F)*	*Pedagogies for Application (PK-A)*	*Tools for Facilitation (TK-F)*	*Tools for Application (TK-A)*
Algorithmic Foundation	Computer Science (AI training), Mathematics (classification), Logic (decision trees)	Adult learner engagement through game-based learning Online interactive facilitation, Scaffolded algorithmic thinking	Plan age-appropriate AI concepts Introduce logical thinking patterns, Connect familiar games to algorithms	• Code.org sorting games • Synchronous conferencing software • Screen sharing for demonstrations	• Simple sorting activities • Physical manipulatives • Basic decision tree templates
Geometric Exploration	Mathematics (geometric properties), Computer Science (feature identification), Logic (pattern recognition)	Guided discovery learning Individual digital worksheets, Structured comparison activities	Facilitate student shape exploration Teach geometric vocabulary, Guide pattern identification	• Google Slides with activity sheets • Individual digital handouts • Online geometric shape libraries	• Physical shape sets • Hands-on sorting, Comparison
Feature Analysis	Mathematics (property classification), Computer Science (algorithm logic), Systems Thinking (hierarchical organization)	Demonstration of matrix creation Adult-level technical explanation, Collaborative feature identification	Teach systematic observation Scaffold logical organization, Guide student categorization	• Digital presentation tools • Interactive matrix templates • Real-time collaborative editing	• Handouts with data tables • Graphic organizers or paper-based drawing
Matrix Creation	Mathematics (systematic classification), Computer Science (supervised learning), Logic (algorithmic representation)	Hands-on algorithm development Adult technical skill building, Peer collaboration and review	Guide student algorithm creation Teach step-by-step procedures, Support logical reasoning	• Advanced spreadsheet tools • Digital matrix builders • Online collaboration platforms	• Basic table creation tools • Paper-based matrices • Hands-on organizational charts

Conclusion

The E-TPACK framework presented in this chapter addresses potential shortcomings in applying TPACK (Mishra & Koehler, 2006) in online PD contexts preparing educators for STEM integration while making the dual-layer and interdisciplinary in STEM explicit. The framework's practical applicability has been demonstrated through implementation examples (Sung & Gunpinar, 2025) showing that authentic STEM connections can be fostered across varied content domains while maintaining accessibility, regardless of technological infrastructure.

The framework's dual-layer approach specifically addresses the complex pedagogical demands of online PD, where PD providers must simultaneously master interdisciplinary content integration while preparing educators for diverse K-12 classroom implementation contexts (Borko, 2004). The COVID-19 Global Spread Analysis Project and AI Literacy Through Geometry Classification examples showed how meaningful STEM integration can occur across diverse educational domains – from biological and geographical modeling to mathematical and AI literacy – while ensuring that well-designed online experiences translate into accessible classroom practices, supporting research indicating that well-designed online PD can achieve equivalent results to face-to-face approaches (Fies & Packham, 2021).

For PD organizations or vendors, these results point to the importance of investing in comprehensive programs that build competency across ICK, EPK, and ETK, recognizing that effective competency development requires sustained support rather than brief training experiences (Darling-Hammond et al., 2017; Guskey, 2000). The demonstrated applications show how such programs can accommodate diverse technological contexts – from sophisticated online collaborative tools to unplugged, hands-on materials – ensuring accessibility across varied resource environments while addressing the reality that most K-12 teachers have been trained within single disciplines rather than integrative approaches (Jong et al., 2021; Lo, 2021).

Educational leaders should prioritize authentic integration practices over isolated content delivery, as both application examples provide scalable solutions that work across diverse educational contexts regardless of resource constraints. The framework's emphasis on dual-layer pedagogical design ensures that online PD can simultaneously provide rich learning experiences for educators and prepare them to implement appropriate adaptations in their own K-12 contexts, whether addressing global health challenges or introducing AI concepts, aligned with research emphasizing the importance of contextual knowledge in technology integration (Petko et al., 2025).

In sum, successful online STEM PD implementation requires attention to context-specific adaptation while maintaining framework integrity and investing in adaptation support that helps PD providers modify approaches for their specific contexts while preserving essential integration features. The systematic application tables demonstrate how content, pedagogical strategies, and technological tools can be carefully selected to serve dual purposes: engaging adult learners in sophisticated online experiences while preparing them for diverse classroom implementation scenarios, reflecting the holistic perspective that effective technology integration requires understanding relationships between technological tools, pedagogical approaches, and CK (Schmidt et al., 2009; Voogt et al., 2013).

The framework's grounding in authentic integration principles and experiential learning, as demonstrated through the application examples, provides a foundation for developing educational approaches that maintain accessibility and pedagogical rigor while being able to adapt to evolving STEM education needs. By addressing the dual-layer complexity inherent in online PD, E-TPACK ensures that STEM integration education remains achievable and meaningful across diverse educational contexts through its emphasis on authentic problem-solving and systematic attention to both facilitation and implementation requirements, contributing to the growing body of research validating TPACK's utility across diverse educational contexts (Phillips, 2017; Wang et al., 2018).

KEY TAKEAWAYS

- Dual-Layer Complexity Requires New Framework: Online STEM PD involves teacher educator-to-teacher-to-student relationships that traditional TPACK does not address, necessitating explicit consideration of both online facilitation strategies and classroom implementation needs.
- Three Extended Components Enable Integration: E-TPACK reconceptualizes TPACK through ICK, EPK with PK-F/PK-A, and ETK with TK-F/TK-A to serve dual purposes in online PD contexts.
- Carefully Designed Online PD Can Translate to Accessible Implementation: The framework enables rich, technology-enhanced PD experiences while ensuring teachers can adapt approaches to diverse classroom contexts, from well-resourced to resource-constrained environments.
- Authentic STEM Integration Transcends Disciplinary Boundaries: The two application examples demonstrate how real-world problem-solving contexts (pandemic investigation, AI literacy) create meaningful connections across science, mathematics, technology, and engineering domains rather than treating them as separate subjects.
- Systematic Design Guidance Supports Scalability: The detailed application tables provide concrete guidance for selecting content, pedagogies, and technologies that simultaneously engage adult learners in sophisticated online experiences while preparing them for practical classroom implementation across varied resource levels.

References

Borko, H. (2004). Professional development and teacher learning. *Educational Researcher*, *33*(8), 3–15. https://doi.org/10.3102/0013189X033008003

Chai, C. S., Koh, J.H.L., & Tsai, C. C. (2013). A review of technological pedagogical content knowledge. *Educational Technology & Society*, *16*(2), 31–51. https://www.jstor.org/stable/jeductechsoci.16.2.31

Darling-Hammond, L., Hyler, M. E., & Gardner, M. (2017). *Effective teacher professional development*. Learning Policy Institute. https://eric.ed.gov/?id=ED606743

English, L. D. (2016). STEM education K-12: Perspectives on integration. *International Journal of STEM Education*, *3*(1), 1–8. https://doi.org/10.1186/s40594-016-0036-1

Fies, C., & Packham, C. (2021, March). Transitioning STEM-focused teacher professional development from f2f to online. In *Society for Information Technology & Teacher Education International Conference* (pp. 486–491). Association for the Advancement of Computing in Education (AACE). https://www.learntechlib.org/primary/p/219175/

Guskey, T. R. (2000). *Evaluating professional development* (Vol. 1). Corwin press.

Jong, M. S. Y., Song, Y., Soloway, E., & Norris, C. (2021). Teacher professional development in STEM education. *Educational Technology & Society*, *24*(4), 81–85.

Kelley, T. R., & Knowles, J. G. (2016). A conceptual framework for integrated STEM education. *International Journal of STEM Education*, *3*(1), 1–11. https://doi.org/10.1186/s40594-016-0046-z

Koehler, M. J., & Mishra, P. (2009). What is technological pedagogical content knowledge? *Contemporary Issues in Technology and Teacher Education*, *9*(1), 60–70.

Lo, C. K. (2021). Design principles for effective teacher professional development in integrated STEM education: A systematic review. *Educational Technology & Society*, *24*(4), 136–152. https://www.jstor.org/stable/48629251

Mishra, P., & Koehler, M. J. (2006). Technological pedagogical content knowledge: A framework for teacher knowledge. *Teachers College Record*, *108*(6), 1017–1054. https://doi.org/10.1111/j.1467-9620.2006.00684.x (Original work published 2006).

Petko, D., Mishra, P., & Koehler, M. J. (2025). TPACK in context: An updated model. *Computers and Education Open*, *8*, 100244. https://doi.org/10.1016/j.caeo.2025.100244

Phillips, M. (2017). Processes of practice and identity shaping teachers' TPACK enactment in a community of practice. *Education and Information Technologies*, *22*(4), 1771–1796. https://doi.org/10.1007/s10639-016-9512-y

Rosenberg, J. M., & Koehler, M. J. (2015). Context and technological pedagogical content knowledge (TPACK): A systematic review. *Journal of Research on Technology in Education*, *47*(3), 186–210. https://doi.org/10.1080/15391523.2015.1052663

Schmidt, D. A., Baran, E., Thompson, A. D., Mishra, P., Koehler, M. J., & Shin, T. S. (2009). Technological pedagogical content knowledge (TPACK): The development and validation of an assessment instrument for preservice teachers. *Journal of Research on Technology in Education*, *42*(2), 123–149. https://doi.org/10.1080/15391523.2009.10782544

Sung, W., & Gunpinar, Y. (2025). Bringing AI literacy into online education: Machine learning integration through geometry in K-12 teacher professional development [Manuscript submitted for publication]. School of Education, The University of Texas at Tyler.

Voogt, J., Fisser, P., Pareja Roblin, N., Tondeur, J., & van Braak, J. (2013). Technological pedagogical content knowledge–a review of the literature. *Journal of Computer Assisted Learning*, *29*(2), 109–121. https://doi.org/10.1111/j.1365-2729.2012.00487.x

Wang, W., Schmidt-Crawford, D., & Jin, Y. (2018). Preservice teachers' TPACK development: A review of literature. *Journal of Digital Learning in Teacher Education*, *34*(4), 234–258. https://doi.org/10.1080/21532974.2018.1498039

SECTION II

Support Materials to Design Equitable Online Continuing Professional Development for STEM Instruction

12

GENDER-EQUITABLE ONLINE STEM TEACHING

Evidence-Based Rubrics for Inclusive Instruction

Siri Anderson and Barbara Billington

Background

Online education presents both challenges and opportunities for addressing gender disparities in science, technology, engineering, and mathematics (STEM) fields. That is, virtual platforms allow for delivering scalable programs for teacher education, but they require thoughtful effort to build teachers' capacity for gender-equitable STEM instruction. Likewise, online learning environments provide more flexible and accessible pathways for underrepresented students to gain STEM acumen but need to be intentionally designed to integrate inclusion and engagement strategies that work for all learners.

In this chapter, we present evidence-based practices for gender-equitable online STEM teaching, strategically articulated in rubrics so that teachers can identify where their practices currently lie and choose areas for improvement. Our goal is to support educators in creating inclusive learning environments that enhance gender equity, achievements, and persistence in the STEM fields. This rubric was developed as part of a four-year National Science Foundation (NSF) grant (#1513060), which researched the student-level impacts of training STEM teachers in strategies known to engage and support female-identified (hereafter girl) students.

Teachers who used these rubrics changed their practices in important ways, and their female students developed stronger STEM identities, increasing their likelihood of persisting in pursuit of future STEM careers (Twin Cities Public Television [TPT], 2021). That is, teachers who used these strategies were more likely to foster their girls' ability to "see" themselves as the kind of person who can (and does) succeed in STEM courses.

Teachers self-reported their activities, submitted evidence of their changes and the outcomes they noticed as a result, and were observed by evaluators. Davis and Davey (2019, 2021), the evaluators, observed that teachers had:

- created more collaborative, social, and community-oriented learning spaces;
- provided opportunities for students to learn from diverse role models;
- increased their use of STEM examples that connected to their students' lives;

DOI: 10.4324/9781003530343-14

- directly supported their female students in solving problems using STEM practices;
- empowered their female students to embrace struggle and overcome perceived challenges in STEM activities; and
- helped their students to identify and challenge STEM stereotypes.

The rubric provides a useful scaffold for teachers wanting to improve their ability to meet diverse students' learning needs. When teachers complete a self-assessment using the rubric, they realize that they do not need to change everything they are doing in order to improve their success with girls (of all backgrounds). Rather, they can choose to make a small shift along a continuum of practices they are already doing to improve their impact on girls' engagement, persistence, and outcomes.

This research was built on decades of applied research initiated by the TPT's SciGirls program and the rubrics were created in collaboration with researchers Alicia Santiago, Brenda Britsch, and evaluator Hilarie Davis. Initially, the strategies were based on the SciGirls Seven: How to Engage Girls in STEM model (Billington et al., 2013), but findings synthesized over multiple research projects and literature reviews (2013, 2019) were integrated into the model to craft these rubrics. The current model integrates three foundational characteristics of teacher ontology needed for STEM courses to be effective with students of all genders. The three undergirding beliefs that teachers must cultivate are:

1 The instructor must believe that a person of any identity can succeed in a STEM field.
2 The instructor must appreciate the value of students' internalizing their relationship with the STEM discipline they are teaching.
3 The instructor must recognize the need for culturally responsive practices.

In addition to these three foundational characteristics, any (or all) of the following six additional strategies—which are recognized across the literature as enhancing STEM instructional success for girls (Billington et al., 2019)—can facilitate more equitable success in a teacher's practice.

1 Connecting STEM experiences to girls' lives.
2 Supporting girls as they investigate questions and solve problems using STEM practices.
3 Empowering girls to embrace struggle, overcome challenges, and increase self-confidence in STEM.
4 Encouraging girls to identify and challenge STEM stereotypes.
5 Emphasizing that STEM is collaborative, social, and community-oriented.
6 Most important, providing opportunities for girls to interact with and learn from STEM role models to whom they can relate.

We recommend that teachers begin by noting where in each rubric they feel their practice fits currently, recognizing that they may have marked items in each of the three categories of practice: novice, proficient, and exemplary. The full original rubric is available online from GenderEquitableTeachingStrategies.weebly.com. The website also includes additional resources about the background behind these proven instructional strategies.

Every learning environment is significantly shaped by the teacher's rapport with their students. As the saying goes, "Students will care about what you have to teach, when they know that you care about them." This has been demonstrated to be particularly true for students who have a tenuous connection with the content (Cavanagh et al., 2018; Collie et al., 2024).

Fostering both a trusting teacher-student relationship and real social connections between students improves the likelihood of students persevering in STEM while modeling an affirmative and culturally relevant understanding of the practice of STEM (Calvin et al., 2022; Erete et al., 2016; Williams-Dobosz et al., 2021; Wilton et al., 2024). Instructors of online STEM classes must be intentional in both (a) building student collaboration into their course design (Fisher et al., 2020) and (b) teaching the importance of collaboration among STEM professionals in the field (Williams-Dobosz et al., 2021). We are social learners, and girls are more likely to be drawn to socially motivated activities than males (Su & Rounds, 2015).

When creating online STEM professional development (PD), instructors should consider resources available to model the practices they seek to have teachers replicate in their own courses. For instance, find STEM practitioners who will visit your course virtually and provide examples of the collaborative work they do regularly, set up small groups (based on availability or interests) so participants get to know one another personally, design a pre-course survey to get to know the students who are enrolled in your course so you can personalize your communication with them, and create a video introduction that orients students to the course learning management system (LMS) and allows them to get to know you. Additionally, build instruction mechanisms for communication and collaboration among participants that extend beyond the course, such as teaching online curating tools that students can use to communicate with one another about course content and their experiences in the field.

Preparation for Online Learning

The shift to online teaching requires intentional preparation and support, particularly when working with historically marginalized student populations (Joosten et al., 2021; Kahn et al., 2022). Our original research (NSF #1513060) encompassed various instructional modalities (in-person, online, hybrid, and train-the-trainer). One factor that influenced STEM teachers' experience of the training was their level of preparation for online learning. This makes sense: Reducing the cognitive load of the mechanism of instruction increases learner resources available to learn the content and skills of the intended lessons. For example, we do not expect students who are learning to drive to simultaneously learn the history of land use practices and how these translate into the location of transportation corridors of various types.

Ideally, the learners in an online STEM class master the online learning system and related skills prior to beginning coursework. Therefore, it is important to pre-assess students for dexterity with the tools they are expected to use. Actively teaching the skills reduces the demands of being an online vs. in-person student. For instance, make sure learners have:

- Knowledge of the LMS areas and functions;
- An online curating tool that will help them organize, locate, and log into resources related to their learning;
- A repertoire of skills for multimodal interactions such as text-to-speech, speech-to-text, captioning, slowing down or speeding up videos/podcasts, translating materials into different languages, embedding links, and a way to check their grammar, spelling, and citations.

Below, we present a framework PD providers can use to help teachers create accessible and equitable online learning environments.

Assessment Rubrics for Gender-Equitable Online STEM Teaching

The following rubrics provide concrete tools to assess and develop teachers' implementation of gender-equitable practices in online STEM instruction. Each rubric includes specific indicators across three levels of implementation, Novice, Proficient, and Exemplary, along with practical examples and research-based strategies. The rubrics are organized into three core domains: Learning Environment, Instructional Strategies, and Identity Development. Each addresses technical competencies and pedagogical considerations.

We found success with the following general approach:

- Teachers read through each rubric top to bottom, left to right, highlighting descriptions of their current practice.
- In dialogue with colleagues, teachers identified areas where they saw opportunities for improvement.
- Consulting with the background information on our website, teachers attempted to make one change in their practice every two weeks for eight weeks. As they did, they noted what changes, if any, they perceived in their students' engagement, effort, and outcomes.
- Teachers brought evidence of "what worked," along with frustrations around what did not work, to discuss with peers. As a group, they might consider: How could the change be made simpler? What results were worth replicating? What challenges could they solve together?
- Teachers identified additional support they would like from the facilitators to help them improve their integration of the gender equitable strategies in their own classrooms.

This cycle of praxis (theory-action-reflection), which was teacher-directed and facilitator-supported, provided agency and improved self-efficacy to the teachers, and their female students' outcomes improved as a result.

Identity Development Domain

This category of teacher practice includes the strategies known to be most impactful in achieving gender equity. The category consists of two components: Integration of Role Models and Challenging STEM Stereotypes.

Online databases are available for finding role models who can visit the class virtually or can be accessed by watching short videos focusing on racially and culturally diverse women working across STEM fields. The role model page of the Gender Equitable Teaching Strategies website (https://genderequitableteachingstrategies.weebly.com) built as part of this grant provides links to many free role model resources across all disciplines of STEM.

School leaders can further support this work by developing a database of local near-peer role models who are willing to connect with students in STEM courses. Rather than finding an astronaut (or other "high-flyer"), teachers should focus on bringing in recent graduates who are pursuing a vocational or STEM postsecondary degree. This sort of role model is more likely to convince skeptical students, those who do not already know someone who "looks like them" working in those fields, that they can and should consider pursuing an advanced degree in these fields.

Another approach is to use online tools to pair students one-to-one with a classroom of students at a local postsecondary institution for interaction around a project. This can happen entirely asynchronously and virtually and still inspire students to believe they have the capacity to succeed. This section has two rubrics to consider:

1 Provide opportunities for girls to interact with and learn from diverse STEM role models.
2 Encourage girls to identify and challenge STEM stereotypes.

Integration of STEM Role Models

Research has consistently demonstrated the powerful influence of role models on students' engagement and interest in STEM fields: When students see individuals from similar backgrounds succeeding, their sense of belonging and self-efficacy significantly increases (Hughes et al., 2020; Kessels, 2023). This is particularly salient for students who may internalize stereotypes about who "belongs" in STEM (Adams et al., 2014; Cheryan et al., 2015; Koch et al., 2015). Providing access to diverse role models challenges these stereotypes by showcasing the attainability of STEM careers for individuals with varied identities (González-Pérez et al., 2020; Hughes et al., 2020; O'Brien et al., 2016). Furthermore, the positive impact of role models extends beyond mere representation; active engagement with role models can foster mentorship relationships, providing invaluable guidance and support (Levine et al., 2015). Finally, longitudinal studies indicate that these interventions can have lasting effects on students' educational and career trajectories, influencing their long-term aspirations and persistence in STEM (Leaper, 2015; Weisgram & Diekman, 2017).

The following rubric guides PD providers in assessing and developing teachers' abilities to provide opportunities for girls to interact with and learn from diverse STEM role models (Table 12.1).

Challenging STEM Stereotypes

Challenging stereotypes is a crucial component of empowering girls in STEM courses, as it fosters an environment where they can see themselves as capable and integral to the field (Allen & Eisenhart, 2017; Boucher et al., 2017). Carli et al. (2016) and Carlone et al. (2015) highlight the importance of creating inclusive curricula that reflect the varied backgrounds and experiences of learners, thereby promoting students' sense of belonging in the STEM classroom. Additionally, when girls perceive STEM as inclusive and representative of their identities, their engagement and persistence in these fields increase (Cheryan et al., 2015; Dasgupta & Stout, 2014). Addressing stereotypes directly, through discussions and reflective activities, can significantly alter perceptions and encourage students to challenge the status quo (Civil, 2016; Robnett, 2016). Girls and students of color who have historically been marginalized from STEM participation benefit from challenging the stereotypes that reinforce their exclusion (Kim et al., 2018). Scott et al. (2014) and Verdin et al. (2016) underscore the necessity of integrating these practices into both virtual and in-person settings to maximize their reach and effectiveness. By doing so, educators can help bridge the gender gap, fostering a new generation of confident, empowered girls ready to lead in STEM.

The following rubric guides PD providers in assessing and developing teachers' abilities to identify and challenge STEM stereotypes (Table 12.2).

TABLE 12.1 Provide Opportunities for Girls to Interact With and Learn From Diverse STEM Role Models

<table>
<tr><td colspan="3">Implementation Example:
An online chemistry instructor can have students write about or draw their vision of a "chemist" and then provide opportunities for them to partner with local industry by creating monthly virtual "Coffee with a Chemist" sessions. Participants prepare questions in advance, engage in 30-minute video conversations with professionals, and maintain ongoing connections through moderated discussion boards. The instructor can archive these sessions and create a searchable database of role model interactions for future reference.</td></tr>
<tr><th>Novice</th><th>Proficient</th><th>Exemplary</th></tr>
<tr><td>Instructor recognizes the value of diverse STEM role models.
Instructor includes some STEM role models in course materials.
Online Implementation
Instructor is respectful and encouraging to all students regardless of their identities.
Instructor sets and maintains high standards of interaction for all learners.
Instructor introduces students to online resources that specialize in diverse engagement in STEM fields.</td><td>Instructor integrates balanced representation of diverse role models across materials.
Instructor facilitates direct role model interactions.
Instructor ensures demographic alignment with the student population.
Online Implementation
Instructor uses discussion boards for ongoing mentor interactions.
Instructor utilizes online role model databases.
Instructor incorporates video profiles in lessons.
Instructor shares digital stories of diverse STEM professionals (e.g., Scientist Spotlight Initiative; https://scientistspotlights.org)
Instructor facilitates online mentoring sessions.</td><td>Instructor offers numerous opportunities for diverse role models to be present in the curriculum.
Instructor establishes connections with community agencies and fosters meaningful relationships between students and various near-peer role models.
Instructor celebrates women of varied backgrounds and time periods visibly within the classroom, community, and with special events.
Online Implementation
Students create their own Scientist Spotlight and upload it to the Initiative's website, thereby building up the digital archive for others.
Instructor hosts virtual visits from STEM professionals.
Instructor creates digital galleries of local STEM alumni.</td></tr>
</table>

Learning Environment Domain

Collaborative STEM Learning Communities Online

Girls thrive in collaborative and inclusive environments that offer contexts for learning that are personally relevant and meaningful (Diekman et al., 2015; Leaper, 2015; Sublett, 2020). Fostering communal goals has been shown to enhance interest and retention in STEM courses for females (Boucher et al., 2017; Fuesting et al., 2019). This approach counters stereotypes of STEM as solitary work and helps students develop essential teamwork skills (Clark et al., 2016).

TABLE 12.2 Encourage Girls to Identify and Challenge STEM Stereotypes

Implementation Example: One of the ways we confront stereotypes is by presenting information that contradicts those stereotypes. Ford and Hines (2025) present numerous online resources you can use to represent STEM engagement of black students. In addition, the SciGirls Role Model Profiles (SciGirls, 2025) provides short videos of women in many different STEM careers. A STEM teacher can assign students to watch and reflect on the materials, considering what it means to them as learners.		
Novice	*Proficient*	*Exemplary*
Instructor recognizes the value of engaging students in developing an awareness of the presence of the gender gap in STEM majors, careers, and voices in industry and research. Instructor shares with students the presence of gender disparities in STEM majors, careers, and voices in industry/research. *Online Implementation* Images and stories in the course materials provide a diverse representation of STEM professionals; however, stereotypes are not addressed.	Instructor invites students to share their STEM conceptions of stereotypical roles in small and large group settings. Instructor collects student feedback on their conceptions of the Draw-A-Scientist-Test (or Draw-An-Engineer-Test) and incorporates it into improved gender-equitable teaching strategies. *Online Implementation* Images and stories in the course materials represent the world as it could be, and at least one topic of stereotype is addressed in the content. Students see themselves as STEM professionals regardless of diverse backgrounds.	Instructor collects student feedback on their conceptions of their own identities as STEM people and incorporates the feedback into improved implementation of strategies at least once for each unit taught. Instructor encourages students to challenge the stereotypical roles they see in their schools, communities, and colleges/universities regarding gender inequities to prompt underrepresented students (girls, students of color) to participate and succeed. *Online Implementation* Students create online information designed to challenge stereotypes in a practice or field (infographics, videos, website, presentation shared on social media) to share publicly or within a specific context (such as younger students).

Research shows student trust of the instructor (Cavanagh et al., 2018) and collaboration opportunities significantly impact student engagement and learning outcomes (Çakır et al., 2017; Capobianco et al., 2015; Ismailov & Chiu, 2022; Robnett, 2013; Wilton et al., 2024). In online environments, structuring collaborative activities requires intentional design (Fisher et al., 2020; Lin & Nixon, 2024).

The following rubric guides PD providers in assessing and developing teachers' abilities to create collaborative online STEM learning communities (Table 12.3).

TABLE 12.3 Emphasis on STEM as Collaborative, Social, and Community-Oriented

<table>
<tr><td colspan="3">Implementation Example:
A computer science teacher explains pair-programming to students and then has a female software engineer from a local company virtually drop into the class to explain why she always has her programmers do pair-programming and how this works to improve user experience and reduce bugs in the code. Students are encouraged to ask her questions about how she got into this role, what she likes most about it, and how it serves her goals in life. Later, the class reads an engineering design problem from engineergirl.org that describes how engineers' work is used to make the world a better place. The class is broken into random groups to approach the problem collaboratively.</td></tr>
<tr><td>Novice</td><td>Proficient</td><td>Exemplary</td></tr>
<tr><td>Instructor knows student names and invites their input on course elements.
Instructor is aware of the efficacy of using diverse forms of lesson activities and interactions to facilitate student learning.
Instructor facilitates student interactions within the classroom/learning space consistently.
Instructor is comfortable managing students working in small groups.
Instructor manages interpersonal conflict and disruptive behavior prosocially.
Online Implementation
Students work in breakout rooms or discussion boards with a small group of learners they get to know and trust.
Instructor checks in with students about any concerns to promote positive relationships.
Instructor creates a safe online learning environment, and students are willing to take academic risks.</td><td>Instructor uses self-study and peer-review to increase the use of active-learning practices.
Instructor integrates accountable talk to facilitate students' success in peer groups.
Instructor aligns course materials with student backgrounds to increase relevance and engagement.
Online Implementation
Games are played to rehearse new content.
Student work groups meet synchronously weekly or bi-weekly to work on assigned problems.
Discussion board prompts foster genuine dialogue, debate, role-plays, or personalized experiences shared by students.
Students self- and peer-assess some assignments with rubrics established in advance of doing the work.</td><td>Students consistently spend learning time in peer groups or with outside mentors to develop their competencies in course standards and skills.
Instructor incorporates student voice into the design of the course.
Instructor uses formative feedback mechanisms to actively improve lesson direction, activities, and outcomes.
Online Implementation
Students and student groups lead online class sessions.
Instructor uses formative assessments to drive instructional decision-making.
Students take responsibility for creating positive relationships that contribute to learning.
Instructor creates virtual meetings (or shows video interviews) with diverse people working in the field, who share stories of how they got to where they are and examples of the collaborative nature of their work.</td></tr>
</table>

Instructional Strategies Domain

In the traditional college experience, courses commonly include three grades per semester—a midterm, a project/paper, and a final exam. This format does not work well in online courses. Students benefit from online courses that are highly structured (Chen et al., 2018; Fisher et al., 2020; Gavassa et al., 2019; Joosten et al., 2021; Sikora et al., 2022). Students in online learning need active learning time, retrieval practice, worked examples, applied problem-solving, teamwork, and formative feedback on their progress (Brown et al., 2021; Kahn et al., 2022; Sandrone et al., 2021).

This section includes three focus areas:

1 Support Girls as They Investigate Questions and Solve Problems Using STEM Practices
2 Connect STEM Experiences to Girls' Lives
3 Empower Girls to Embrace Struggle, Overcome Challenges, and Increase Self-Confidence in STEM

Student-Centered Solving and Investigation

Research consistently demonstrates that effective STEM education requires active student engagement in authentic scientific and engineering practices. This is particularly important for girls and students from underrepresented groups, who benefit from opportunities to actively and collaboratively participate in STEM inquiry rather than passively receiving information (Buchholz et al., 2014; McLean et al., 2020; Munley & Rossiter, 2013). In online learning environments, structured problem-solving experiences must be intentionally designed to promote student agency and collaboration (Chen et al., 2018; Fisher et al., 2020). PD providers should help teachers move beyond "delivering content" to "facilitating student-driven investigation," where learners engage with the full complement of science and engineering practices outlined by the National Research Council (2013). When teachers, or role models, support girls in developing their own questions, designing investigations, and constructing explanations, they help them build their identities as capable STEM practitioners (González-Pérez et al., 2020). Online learning platforms offer unique affordances for these experiences, including access to simulations, collaborative tools, and real-time data visualization (Gavassa et al., 2019; Sikora et al., 2022).

The following rubric guides PD providers in assessing and developing teachers' capacity to support student-centered investigations in online STEM courses, with particular attention to creating inclusive environments where all students, especially girls, engage deeply with STEM practices (Civil, 2016; Joosten et al., 2021; Kim, 2016; Russo, 2023; Scott & White, 2013) (Table 12.4).

Connecting STEM to Students' Lives and Cultural Backgrounds

Effective STEM instruction necessitates a pedagogical shift toward connecting content with students' lived experiences and cultural backgrounds. Recognizing that students enter classrooms with diverse knowledge and perspectives, educators should integrate these assets into the curriculum (Cervantes-Soon, 2016; Civil, 2016; Gay, 2013). This approach not only enhances engagement but also fosters a deeper understanding of STEM concepts by

TABLE 12.4 Support Girls as They Investigate Questions and Solve Problems Using STEM Practices

Implementation Example: An online math teacher has established groups of students who collaborate on problem-solving methods. Students are given story problems that represent realistic contexts from their own lives and asked to come up with three different ways to solve the problems mathematically. Students have access to sentence stems and accountable talk guidelines to shape their discussions.		
Novice	*Proficient*	*Exemplary*
Instructor recognizes the value of promoting students' creativity in STEM. Instructor allows for flexible student choice in presenting content during the semester. Instructor is aware that activities and practices that promote critical thinking increase students' critical thinking capacities. *Online Implementation* Students utilize the Next Generation Science Standards (NGSS) practices with some frequency; however, typically in isolation rather than the full complement of eight science and engineering practices. Instructor engages students in authentic STEM problem-solving in virtual or online spaces utilizing the NGSS science and engineering practices. Students use interactive learning objects, video-embedded quizzes, or conversational AI to test their developing understanding of new concepts.	Review of curriculum materials demonstrates frequent opportunities for student choice in presenting data/ understanding of STEM concepts utilizing some of the NGSS science and engineering practices. Periodically (e.g., twice a month), students have an opportunity to complete open-ended assignments in which they apply an array of problem-solving methods and transfer their learning to new situations using the skill sets of scientists and engineers. Review of curriculum materials demonstrates daily opportunities for (and modeling of) students' engagement in critical thinking about STEM content in writing or in discussion. *Online Implementation* Students utilize some of the NGSS science and engineering practices, but not yet the full complement of practices in one project/problem.	Instead of assignments with "one right answer," the instructor points to many assignments in which the goals align with creative problem solving, provide affordances for coming to solutions, and encourage authentic learning. In discussions, students are encouraged to work through their initial ideas on a topic, use resources to find answers independently, and bring ideas that are genuinely their own into the conversation. Instructor creates a challenging learning environment that exemplifies critical thinking capacities of students discussing STEM concepts, practices, and phenomenon. *Online Implementation* Students engage in challenging problem-solving (independently and collaboratively), construct their own knowledge, and demonstrate enthusiasm and effort in completing work utilizing the eight NGSS science and engineering practices in complementary ways.

illustrating their relevance to students' daily lives (Bonner & Dorneich, 2016; Hughes et al., 2020). Furthermore, culturally responsive STEM education acknowledges and validates students' cultural identities, promoting a sense of belonging and empowerment within the learning environment (Cobian et al., 2024; Hubert, 2014; Rainey et al., 2018). By weaving in real-world applications and culturally relevant contexts, STEM instruction can become more accessible and meaningful, ultimately leading to improved student outcomes and a more inclusive STEM community (Boucher et al., 2017; Ladson-Billings, 2008; Verdin et al., 2016).

The following rubric guides PD providers in assessing and developing teachers' abilities connect STEM experiences to students' lives and cultural backgrounds (Table 12.5).

TABLE 12.5 Connect STEM Experiences to Girls' Lives

Implementation Example: In a physics class, students are taught about the impact of the fashion industry on the environment. In small groups, students generate lists of all the places where energy transfer happens in the life cycle of a piece of clothing. Students are divided into three problem-solving groups around the following questions: (a) How can we generate energy to power clothing (e.g., solar panels, kinetic energy)? (b) How can we regulate temperature in clothing to reduce energy consumption (e.g., heat transfer, insulation)? (c) How can we incorporate technology to make clothing more efficient (e.g., smart textiles, sensors)? And (d) What can be done to reduce the amount of waste that is created from one clothing item's production and life cycle? Each group presents its findings to the class with explanations that include at least two of the following: energy transfer, kinetic energy, thermal energy, and energy conservation.		
Novice	*Proficient*	*Exemplary*
Instructional strategies and learning activities are informed by knowledge of students' interests and skills. Instructor recognizes the value of cultural awareness and making learning experiences relevant to students' lives. Intentionally chooses culturally relevant practices and curriculum materials to engage learners. Instructor is aware that students benefit from increased cultural relevance in all STEM experiences. *Online Implementation* Instructor is aware of implicit bias present in online resources that perpetuate inequities (for women and diverse students). Instructor shares examples of professionals in STEM careers who represent the diverse student body in their online course.	Instructional strategies and learning activities are informed by knowledge of students' interests and skills as well as cultural backgrounds, and language proficiency. Curriculum materials and visuals demonstrate examples of cultural awareness of the students in each unit. Instructor invites students to share their experiences related to the STEM content they are learning, to provide opportunities for student voice to be included. Instructor actively explores personal values, beliefs, strengths, and challenges in relation to culture (gender, race, economic status). *Online Implementation* Instructor provides relevant STEM examples and references for students in terms of their home communities in course materials.	Students' experiences are consistently connected to their culture via activities and community-based experiences. Students' autobiographical information is actively incorporated into the course. Instructor establishes active connections between STEM content and personal interest/ experiences. Instructor implements numerous strategies that engage their diverse learners to see the relevance in their STEM content. *Online Implementation* Instructor establishes an environment in which students respect and affirm their own and others' differences and are supported to share and explore differences and similarities related to their cultural background and identity. Instructor embeds relevant STEM examples for students and their home communities in most course materials.

Empowering Students in STEM Learning

Students benefit from explicit instruction in neuroplasticity and the importance of a growth mindset, where intelligence is recognized as malleable rather than fixed (Dweck, 2000). Societal messages and stereotypes are directly addressed and challenged within the course curriculum (Halpern et al., 2007; Kim et al., 2007). Feedback focuses on effort and strategy, rather than innate ability, to foster persistence and achievement (Blackwell et al., 2007; Mueller & Dweck, 1998; Yeager et al., 2021). Challenging stereotypes, promoting a growth mindset, and emphasizing the value of hard work and learning empowers girls and other historically marginalized populations to thrive in STEM. Nation et al. (2019) use the compelling term of facilitating students becoming *STEMinists,* recognizing science as a practice of pushing through difficulty. The following rubric guides PD providers in assessing and developing teachers' abilities to empower girls to embrace struggle, overcome challenges, and increase self-confidence in STEM (Table 12.6).

TABLE 12.6 Empower Girls to Embrace Struggle, Overcome Challenges, and Increase Self-Confidence in STEM

Implementation Example: A physics teacher has set up a discussion board where students can privately communicate with him. He notices that several students have asked about the current assignment, so he creates a video where he thanks the students for their questions and provides a worked example of a similar problem in a familiar context for the students. After sharing the video with students individually, he offers to log onto their next synchronous class session 15 minutes early if they would like to join other students who had similar questions to discuss the assignment.		
Novice	*Proficient*	*Exemplary*
Instructor can articulate the difference between a growth and a fixed mindset within their discipline. Instructor implements a strategy to provide equitable feedback to learners (such as asking questions equally often to all students, providing similar wait time to students regardless of background). Instructor shows genuine respect for students and belief in their capability. Instructor communicates clear and specific expectations. *Online Implementation* Instructor ensures students are prepared for online learning	Instructor uses blind review of formative and summative assessments to self-assess the ways in which feedback is provided to learners to further their understanding. Instructor's tone is respectful, and their demeanor is encouraging of high standards for all students. Students see failure as a positive learning experience. *Online Implementation* Instructor ensures that online interactions and instructional outcomes convey high expectations for all students and	Anonymous student feedback on course/instructor shows that all students find the instructor to be encouraging and supportive. In small group dialogues, it is easy to hear/see that students know how to: talk to one another in academically meaningful ways; ask and answer questions of one another; and treat one another with respect. Instructor has a positive attitude toward the potential of all students to succeed. When a student isn't succeeding, the instructor reconsiders how they could differently

(Continued)

TABLE 12.6 (Continued)

Novice	*Proficient*	*Exemplary*
demands prior to beginning of course. Instructor ensures materials are compliant with the Americans with Disabilities Act (ADA). Instructor shares a specific story that underlines their belief in the ability of all students to succeed, as part of a video introduction to a course or early module.	scaffolds are put in place for struggling students to improve their performance such as an ask-and-answer-a-question discussion area, and online office hour—both virtual and live. Instructor redesigns assignments to afford all learners ample opportunities to receive formative feedback from the instructor on learning and improve outcomes (resubmission of improved work is accepted and evaluated).	approach the content or pedagogy to afford success. *Online Implementation* Instructor consistently and clearly models how students can master challenging material and meet learning goals through effective effort. Instructor consistently enforces expectations for student effort, behavior, and work with special recognition as needed. Instructor automates personalized check-in emails with students who score 70% or lower on assignments to see if they would like additional support.

Discussion

Teaching is an art, a science, and a complex web of interpersonal knowledge, skills, and relationships. Remembering that we teach people, not content, can help underscore for STEM faculty the primacy of focusing on the lived experiences and outcomes of their students. Ensuring everyone feels a sense of belonging in the classroom and within the discipline increases the likelihood of students persisting and doing well in the subject matter (Rainey et al., 2018). STEM teachers' sense of belonging in STEM can be fostered by allowing them to self-identify their instructional strategies on the rubrics embedded in this chapter and then allowing their own interests to drive their direction of growth. This process of self-reflection through the rubrics affords teachers the opportunity to develop their instructional practices toward more equitable methods and educating all students in their online classrooms.

Online STEM instructors can use readily available materials, instructional strategies, assignment structures, and feedback mechanisms to create classes that support all learners' pathways to success. As revealed in the above rubrics, small changes in the learning environment and instructional strategies, with attention to developing students' STEM identity, can lead to increased access and success for all students in online STEM courses. When instructors move from awareness of these strategies to consistently providing students with access to active learning environments, opportunities to interact with peers and role models, as well as solving relevant STEM problems in small groups, instructors move beyond Novice to Proficient and Exemplary online STEM instruction through an iterative process of reflection and praxis.

KEY TAKEAWAYS

- Small identifiable changes in practice will improve STEM teachers' efficacy with girls in their classes.
- Teachers can make progress by identifying where their current practices lie on a continuum and choosing where to make improvements.
- Online and in-person STEM classes that integrate near-peer role models, challenge stereotypes, and create collaborative learning environments will inspire and engage girls.
- STEM classes should model the practices of real STEM disciplinary work, including collaborative and social problem-solving, allowing for struggle, and celebrating challenges/failures.

References

Adams, J. D., Gupta, P., & Cotumaccio, A. (2014). Long-term participants: A museum program enhances girls' STEM interest, motivation, and persistence. *Afterschool Matters*, *20*, 13–20.

Allen, C. D., & Eisenhart, M. (2017). Fighting for desired versions of a future self: How young women negotiated STEM-related identities in the discursive landscape of educational opportunity. *Journal of the Learning Sciences*, *26*(3), 407–436. https://doi.org/10.1080/10508406.2017.1294985

Blackwell, L. S., Trzesniewski, K. H., & Dweck, C. (2007). Implicit theories of intelligence predict achievement across an adolescent transition: A longitudinal study and an intervention. *Child Development*, *78*(1), 246–263. https://doi.org/10.1111/j.1467-8624.2007.00995.x

Billington, B., Britsch, B., Karl, R., Carter, S., Freese, J., & Regalla, L. (2013). SciGirls Seven: How to engage girls in STEM. *Twin Cities Public Television.*

Billington, B., Britsch, B., Santiago, A., Schellinger, J., Carter, S., & Becker, N. (2019). SciGirls strategies: How to engage girls in STEM. SciGirls connect. *TPT National Publications.*

Bonner, D., & Dorneich, M. (2016, September). Developing game-based learning requirements to increase female middle school students' interest in computer science. In *Proceedings of the Human Factors and Ergonomics Society annual meeting* (Vol. 60, No. 1, pp. 380–384). SAGE Publications. https://doi.org/10.1177/1541931213601086

Boucher, K. L., Fuesting, M. A., Diekman, A. B., & Murphy, M. C. (2017). Can I work with and help others in this field? How communal goals influence interest and participation in STEM fields. *Frontiers in Psychology*, *8*, 901. https://doi.org/10.3389/fpsyg.2017.00901

Brown, T., Rongerude, J., Leonard, B., & Merrick, L. C. (2021). Best practices for online team-based learning: Strengthening teams through formative peer evaluation. *New Directions for Teaching and Learning*, *2021*(165), 53–64. https://doi.org/10.1002/tl.20436

Buchholz, B., Shively, K., Peppler, K., & Wohlwend, K. (2014). Hands on, hands off: Gendered access in crafting and electronics practices. *Mind, Culture, and Activity*, *21*(4), 278–297. https://doi.org/10.1080/10749039.2014.939762

Çakır, N. A., Gass, A., Foster, A., & Lee, F. J. (2017). Development of a game-design workshop to promote young girls' interest towards computing through identity exploration. *Computers & Education*, *108*, 115–130. https://doi.org/10.1016/j.compedu.2017.02.002

Calvin, J. J., Ondry, J. C., Dahl, J. C., Sedlak, A. B., McKeown-Green, A., Wang, X., Crook, M. F., Gleason, S. P., Hauwiller, M. R., Baranger, A. M., & Alivisatos, A. P. (2022). Research group-led undergraduate research program: Analyzing and improving a versatile springboard for first-year undergraduates. *Journal of Chemical Education*, *99*(2), 799–809. https://doi.org/10.1021/acs.jchemed.1c01065

Capobianco, B. M., Ji, H. Y., & French, B. F. (2015). Effects of engineering design-based science on elementary school science students' engineering identity development across gender and grade. *Research in Science Education*, *45*(2), 275–292. https://doi.org/10.1007/s11165-014-9422-1

Carli, L. L., Alawa, L., Lee, Y., Zhao, B., & Kim, E. (2016). Stereotypes about gender and science: Women ≠ scientists. *Psychology of Women Quarterly*, *40*(22), 244–260. https://doi.org/10.1177/0361684315622645

Carlone, H. B., Johnson, A., & Scott, C. M. (2015). Agency amidst formidable structures: How girls perform gender in science class. *Journal of Research in Science Teaching*, *52*(4), 474–488. https://doi.org/10.1002/tea.21224

Cavanagh, A. J., Chen, X., Bathgate, M., Frederick, J., Hanauer, D. I., & Graham, M. J. (2018). Trust, growth mindset, and student commitment to active learning in a college science course. *CBE—Life Sciences Education*, *17*(1), ar10. https://doi.org/10.1187/cbe.17-06-0107

Cervantes-Soon, C. G. (2016). Mujeres truchas: Urban girls redefining smartness in a dystopic global south. *Race, Ethnicity, and Education*, *19*(6), 1209–1222. https://doi.org/10.1080/13613324.2016.1168543

Chen, B., Bastedo, K., & Howard, W. (2018). Exploring design elements for online STEM courses: Active learning, engagement & assessment design. *Online Learning*, *22*(2), 59–75. https://doi.org/10.24059/olj.v22i2.1369

Cheryan, S., Master, A., & Meltzoff, A. N. (2015). Cultural stereotypes as gatekeepers: Increasing girls' interest in computer science and engineering by diversifying stereotypes. *Frontiers in Psychology*, *6*, 49. https://doi.org/10.3389/fpsyg.2015.00049

Civil, M. (2016). STEM learning research through a Funds of Knowledge lens. *Cultural Studies of Science Education*, *11*(1), 41–59. https://doi.org/10.1007/s11422-014-9648-2

Clark, E. K., Fuesting, M. A., & Diekman, A. B. (2016). Enhancing interest in science: Exemplars as cues to communal affordances of science. *Journal of Applied Social Psychology* *46*(11), 641–654. https://doi.org/10.1111/jasp.12392

Cobian, K. P., Hurtado, S., Romero, A. L., & Gutzwa, J. A. (2024). Enacting inclusive science: Culturally responsive higher education practices in science, technology, engineering, mathematics, and medicine (STEMM). *PLOS*, *19*(1), e0293953. https://doi.org/10.1371/journal.pone.0293953

Collie, R. J., Martin, A. J., Flesken, A., & McCourt, B. (2024). Personal agency among students from low socio-economic backgrounds: An examination of student profiles, perceived teaching support, and achievement. *Social Psychology of Education*, *27*(4), 1705–1736. https://doi.org/10.1007/s11218-023-09881-0

Dasgupta, N., & Stout, J. G. (2014). Girls and women in science, technology, engineering, and mathematics: STEMing the tide and broadening participation in STEM careers. *Policy Insights from the Behavioral and Brain Sciences*, *1*(1), 21–29. https://doi.org/10.1177/2372732214549471

Davis, H., & Davey, B. (2021, May). *SciGirls strategies implementation evaluation report*. TLC Inc.

Davis, H. B., & Davey, B. T. (2019). *SciGirls strategies gender equitable teaching and advising strategies (GETAS) training evaluation report*. TLC Inc.

Diekman, A. B., Weisgram, E. S., & Belanger, A. L. (2015). New routes to recruiting and retaining women in STEM: Policy implications of a communal goal congruity perspective. *Social Issues and Policy Review*, *9*(1), 52–88. https://doi.org/10.1111/sipr.12010

Dweck, C. S. (2000). *Self-theories: Their role in motivation, personality and development*. Psychology Press. https://doi.org/10.4324/9781315783048

Erete, S., Pinkard, N., Martin, C. K., & Sandherr, J. (2016, August). Exploring the use of interactive narratives to engage inner-city girls in computational activities. In J. Payton (Ed.), *Research on equity and sustained participation in engineering, computing, and technology* (RESPECT) (pp. 1–4). IEEE. https://doi.org/10.1109/RESPECT.2016.7836168

Fisher, D., Frey, N., & Hattie, J. (2020). *The distance learning playbook, grades K-12: Teaching for engagement and impact in any setting*. Corwin Press. https://doi.org/10.4135/9781071838532

Ford, D. Y., & Hines, E. M. (2025). "We love STEM too": Fostering STEM engagement among gifted and talented black students using contemporary young STEM Black models. *Gifted Child Today*, *48*(1), 73–78. https://doi.org/10.1177/10762175241287877

Fuesting, M. A., Diekman, A. B., Boucher, K. L., Murphy, M. C., Manson, D. L., & Safer, B. L. (2019). Growing STEM: Perceived faculty mindset as an indicator of communal affordances in STEM. *Journal of Personality and Social Psychology*, *117*(2), 260. https://doi.org/10.1037/pspa0000154

Gavassa, S., Benabentos, R., Kravec, M., Collins, T., & Eddy, S. (2019). Closing the achievement gap in a large introductory course by balancing reduced in-person contact with increased course structure. *CBE—Life Sciences Education*, *18*(1). https://doi.org/10.1187/cbe.18-08-0153

Gay, G. (2013). Teaching to and through cultural diversity. *Curriculum Inquiry*, *43*(1), 48–70.

González-Pérez, S., Mateos de Cabo, R., & Sáinz, M. (2020). Girls in STEM: Is it a female role-model thing? *Frontiers in Psychology*, *11*, 564148. https://doi.org/10.3389/fpsyg.2020.02204

Halpern, D., Aronson, J., Reimer, N., Simpkins, S., Star, J., & Wentzel, K. (2007). Encouraging girls in math and science (NCER 2007-2003). U.S. Department of Education, National Center for Education Research, Institute of Education Sciences.

Hubert, T. L. (2014). Learners of mathematics: High school students' perspectives of culturally relevant mathematics pedagogy. *Journal of African American Studies*, *18*(3), 324–336. https://doi.org/10.1007/s12111-013-9273-2

Hughes, R., Schellinger, J., Billington, B., Britsch, B., & Santiago, A. (2020). A summary of effective gender equitable teaching practices in informal STEM education spaces. *Journal of STEM Outreach*, *3*(1), 1–9. https://doi.org/10.15695/jstem/v3i1.16

Ismailov, M., & Chiu, T. K. (2022). Catering to inclusion and diversity with universal design for learning in asynchronous online education: A self-determination theory perspective. *Frontiers in Psychology*, *13*, 819884. https://doi.org/10.3389/fpsyg.2022.819884

Joosten, T., Harness, L., Poulin, R., Davis, V., & Baker, M. (2021). *Research review: Educational technologies and their impact on student success for racial and ethnic groups of interest*. WICHE Cooperative for Educational Technologies (WCET). https://wcet.wiche.edu/resources/research-review-educational-technologies-and-their-impact-on-student-success-for-racial-and-ethnic-groups-of-interest/

Kahn, S., Vertesi, J., Adriaenssens, S., Byeon, J., Fixdal, M., Godfrey, K., Lumbroso, J., & Wagoner, K. (2022). The impact of online STEM teaching and learning during COVID-19 on underrepresented students' self-efficacy and motivation. *Journal of College Science Teaching*, *51*(6), 6–15. https://doi.org/10.1080/0047231X.2022.12315644

Kessels, U. (2023). Gender identity and academic engagement: How the perceived fit between students' gender and school work explains gendered pathways in education. In *The Routledge International Handbook of Gender Beliefs, Stereotype Threat, and Teacher Expectations* (pp. 44–55). Routledge.

Kim, A. Y., Sinatra, G. M., & Seyranian, V. (2018). Developing a STEM identity among young women: A social identity perspective. *Review of Educational Research*, *88*(4), 589–625. https://doi.org/10.3102/0034654318779957

Kim, H. (2016). Inquiry-based science and technology enrichment program for middle school-aged female students. *Journal of Science Education & Technology*, *25*(2). https://doi.org/10.1007/s10956-015-9584-2

Kim, Y., Wei, Q., Xu, B., Ko, Y., & Ilieva, V. (2007). MathGirls: Toward developing girls' positive attitude and self-efficacy through pedagogical agents. In R. Luckin, K. R. Koedinger, & J. Greer (Eds.), *Artificial intelligence in education: Building technology rich learning contexts that work* (pp. 119–126). IOS Press.

Koch, M., Lundh, P., & Harris, C. J. (2015). Investigating STEM support and persistence among urban teenage African American and Latina girls across settings. *Urban Education*, *54*(2), 243–273. https://doi.org/10.1177/0042085915618708

Ladson-Billings, G. (2008). "Yes, but how do we do it?": Practicing culturally relevant pedagogy. In W. Ayers (Ed.), *City kids, city schools: More reports from the front row* (pp. 162–177). W. W. Norton.

Leaper, C. (2015). Do I belong?: Gender, peer groups, and STEM achievement. *International Journal of Gender, Science and Technology*, *7*(2), 166–179.

Levine, M., Serio, N., Radaram, B., Chaudhuri, S., & Talbert, W. (2015). Addressing the STEM gender gap by designing and implementing an educational outreach chemistry camp for middle school girls. *Journal of Chemical Education*, *92*(10), 1639–1644. https://doi.org/10.1021/ed500945g

Lin, Y., & Nixon, N. (2024, July). STEM pathways in a global online course: Are male and female learners motivated the same? In *Proceedings of the eleventh ACM conference on learning@ scale* (pp. 243–249). ACM. https://doi.org/10.1145/3657604.3662045

McLean, M., Nation, J. M., & Spina, A. (2020). The importance of collaborative design for narrowing the gender gap in engineering: An analysis of engineering identity development in elementary students. *Journal of Pre-College Engineering Education Research* (J-PEER), *10*(2), 2. https://doi.org/10.7771/2157-9288.1210

Mueller, C. M., & Dweck, C. S. (1998). Praise for intelligence can undermine children's motivation and performance. *Journal of Personality and Social Psychology*, *75*(1), 33–52.

Munley, M. E., & Rossiter, C. (2013). *Girls, equity and STEM in informal learning settings: A review of literature*. National Research Council.

Nation, J. M., Harlow, D., Arya, D. J., & Longtin, M. (2019). Being and becoming scientists: Design-based STEM programming for girls. *Afterschool Matters*, *29*, 36–44.
National Research Council. (2013). *Next Generation Science Standards: For states, by states*. The National Academies Press. https://doi.org/10.17226/18290
O'Brien, L. T., Hitti, A., Shaffer, E., Van Camp, A. R., Henry, D., & Gilbert, P. N. (2016). Improving girls' sense of fit in science increasing the impact of role models. *Social Psychological and Personality Science*, *8*(3), 301–309. https://doi.org/10.1177/1948550616671997
Rainey, K., Dancy, M., Mickelson, R., Stearns, E., & Moller, S. (2018). Race and gender differences in how sense of belonging influences decisions to major in STEM. *International Journal of STEM Education*, *5*, 1–14. https://doi.org/10.1186/s40594-018-0115-6
Robnett, R. (2013). *The role of peer support for girls and women in the stem pipeline: Promoting identification with stem and mitigating the negative effects of sexism* (Doctoral dissertation, University of California, Santa Cruz). https://escholarship.org/uc/item/8nx2r09d
Robnett, R. D. (2016). Gender bias in STEM fields: Variation in prevalence and links to STEM self-concept. *Psychology of Women Quarterly*, *40*(1), 65–79. https://doi.org/10.1177/0361684315596162
Russo, D.M.C. (2023). *A community of inquiry and Hispanic student intent to persist in STEM*. ProQuest LLC. https://www.proquest.com/docview/2776687047?pq-origsite=primo
Sandrone, S., Scott, G., Anderson, W. J., & Musunuru, K. (2021). Active learning-based STEM education for in-person and online learning. *Cell*, *184*(6), 1409–1414. https://doi.org/10.1016/j.cell.2021.01.045
Scott, K., Aist G., & Zhang, X. (2014). Designing a culturally responsive computing curriculum for girls. *International Journal of Gender, Science and Technology*, 6(2), 264–276. https://genderandset.open.ac.uk/index.php/genderandset/article/view/361
Scott, K. A., & White, M. A. (2013). COMPUGIRLS' standpoint: Culturally responsive computing and its effect on girls of color. *Urban Education*, *48*(5), 657–681. https://doi.org/10.1177/0042085913491219
SciGirls. (2025). *SciGirls collect role model videos*. Public Broadcasting System. https://tpt.pbslearningmedia.org/collection/scigirls/
Sikora, A. K., Hall, B., Mills, S., Roberts, R., & Craig, P. (2022). You gotta work, BASIL! Reimagining an established CURE to provide high-quality digital learning experiences that are intentionally equitable, inclusive and accessible for all students. *The FASEB Journal*, *36*(S1). https://doi.org/10.1096/fasebj.2022.36.S1.R3061
Su, R., & Rounds, J. (2015). All STEM fields are not created equal: People and things interests explain gender disparities across STEM fields. *Frontiers in Psychology*, *6*, 189. https://doi.org/10.3389/fpsyg.2015.00189
Sublett, C. (2020). *Distant equity: The promise and pitfalls of online learning for students of color in higher education*. American Council on Education.
Twin Cities Public Television Inc. (TPT). (2021). *SciGirls strategies: Gender equitable teaching practices in career and technical education pathways for high school girls. Final research report*. National Science Foundation #1513060.
Verdin, D., Godwin, A., & Capobianco, B. (2016). Systematic review of the funds of knowledge framework in STEM education. (School of Engineering Education Graduate Student Series. Paper 59). ASEE Annual Conference and Exposition, New Orleans, LA. https://doi.org/10.18260/p.25999
Weisgram, E. S., & Diekman, A. B. (2017). Making STEM "family friendly": The impact of perceiving science careers as family-compatible. *Social Sciences*, *6*(2), 61. https://doi.org/10.3390/socsci6020061
Williams-Dobosz, D., Jeng, A., Azevedo, R. F., Bosch, N., Ray, C., & Perry, M. (2021). Ask for help: Online help-seeking and help-giving as indicators of cognitive and social presence for students underrepresented in chemistry. *Journal of Chemical Education*, *98*(12), 3693–3703. https://doi.org/10.1021/acs.jchemed.1c00839
Wilton, M., Maloy, J., Beaster-Jones, L., Sato, B. K., Lo, S. M., & Grunspan, D. Z. (2024). Instructional influencers: Teaching professors as potential departmental change agents in diversity, equity, and inclusion. *CBE—Life Sciences Education*, *23*(3), ar35. https://doi.org/10.1187/cbe.24-03-0102
Yeager, D. S., Carroll, J. M., Buontempo, J., Cimpian, A., Woody, S., Crosnoe, R., Muller, C., Murray, J., Mhatre, P., Kersting, N., Hulleman, C., Kudym, M., Murphy, M., Duckworth, A. L., Walton, G. M., & Dweck, C. S. (2021). Teacher mindsets help explain where a growth-mindset intervention does and doesn't work. *Psychological Science*, *33*(1), 18–32. https://doi.org/10.1177/09567976211028984

13

OPEN EDUCATIONAL RESOURCES (OERS) AND ONLINE PROFESSIONAL LEARNING NETWORKS (PLNS) IN STEM EDUCATION

Yujin Park, Jewoong Moon, and Hunhui Na

Introduction

Science, technology, engineering, and mathematics (STEM) educators encounter distinct challenges in their teaching due to the fast-changing landscape of STEM. In particular, these teachers face considerable obstacles, including issues related to pedagogy, curriculum, school structure, student needs, assessment methods, and insufficient support for teachers (Lin et al., 2021). Additionally, the constantly evolving technological environment places extra pressure on STEM instructors, whose work is heavily reliant on technology that rapidly changes (Tan et al., 2019). These circumstances call for flexible and accessible professional development resources that can adapt to these swift changes while meeting the varied learning needs of teachers, extending beyond traditional training programs.

Open Educational Resources (OERs) offer a valuable resource to address the professional development needs and challenges faced by STEM teachers in diverse teaching and learning environments. OER includes a broad range of educational materials that are either in the public domain or released under an open license, thereby allowing users to legally and freely copy, utilize, adapt, and redistribute them. This category encompasses textbooks, lecture notes, assignments, as well as videos, animations, and audio files (Hacisalihoglu, 2024). These resources are abundant, customizable, cost-free, and easy to integrate into lesson plans (U.S. Department of Education, 2024).

The demand for accessible, high-quality OER in STEM education is particularly pressing because of the specialized knowledge required and the swift advancements in technology. OER integration in the teaching of STEM subjects is especially relevant because they demand specific content and detailed explanations (U.S. Department of Education, 2024). Further, the versatility of OER empowers teachers engaged in self-directed professional growth to tailor their professional development in a way that suits their specific subject matter, grade levels, and technological skills.

DOI: 10.4324/9781003530343-15

STEM Educators' Need for Professional Development

Research on standard professional development for STEM teachers indicates that most of these programs focus primarily on pedagogical knowledge (PK) and pedagogical content knowledge (PCK). In addition, many programs emphasize technological content knowledge (TCK) and specific technological pedagogical content knowledge (TPCK) (Liu et al., 2022). However, many educators do not have access to ongoing, formal professional development opportunities, making informal learning through OER essential for staying current in their fields and acquiring the TPCK needed for effective STEM teaching (Kelley & Knowles, 2016). The integration of technology in STEM education is not merely supplementary but fundamental to contemporary pedagogical practice. Therefore, as Liu et al. (2022) demonstrated, effective STEM teacher professional development must address both TCK and TPCK alongside traditional pedagogical skills. This technological dimension permeates all aspects of OER utilization, from accessing digital repositories to implementing interactive simulations in classroom instruction.

Current challenges in effective STEM teacher professional development emphasize the need to increase the availability and accessibility of OER for informal learning environments. Teachers indicate that factors aiding their efforts to implement STEM education include collaboration with colleagues, high-quality curricula, support from the district, prior experiences, and effective professional development (Lin et al., 2021). Furthermore, many STEM teachers have entered the profession through alternative qualification programs rather than traditional teacher education programs, leading to gaps in teaching experience and formal training in teaching methods (Zhang et al., 2023). These pedagogical deficiencies underscore the need for continuous support for informal professional development, especially resources and environments that connect theoretical knowledge with practical classroom applications, such as Professional Learning Networks (PLNs). Again, this is where OER can play a significant role.

To fully reap the potential of OER for informal professional development of STEM teachers, several important factors need to be considered. Specifically, there is a need for OER collections that cater to the specific professional learning requirements of STEM educators, focusing on areas such as innovative teaching methods, technology integration, and interdisciplinary strategies (Ma, 2024). By addressing these specific needs through well-designed OER programs, we can enhance teachers' continuous professional growth and ultimately boost student learning outcomes in a technology-driven environment.

In this chapter, we examine OER and PLNs as pathways for the informal professional development of STEM educators, along with practical resources and platforms related to OER.

Use of Open Educational Resources (OERs) in STEM Education

OERs refer to educational materials that are freely available to the public without any cost restrictions, designed for pedagogical use. These resources encompass a broad range of materials, including comprehensive courses, individual components, standalone learning objects, open-access textbooks, streaming videos, assessment tools, and software applications. This wide variety of materials exemplifies the evolving landscape of educational resources, providing educators and learners with diverse tools to enhance teaching and learning practices (Butcher & Moore, 2015). In particular, OER in STEM education helps visualize abstract concepts and improve access to complex topics.

The incorporation of OER into STEM education provides cost-effective, adaptable, and customizable materials that enrich teaching experiences. These resources are especially beneficial in STEM fields, where they encourage innovative teaching methods and broaden access to high-quality educational content. Various studies highlight the ability of OER to enhance educational outcomes and foster active learning environments in teacher professional development contexts. For example, by removing financial barriers through free access to educational materials, OER can help teachers enhance their teaching practices and build competencies in STEM pedagogy (Hacisalihoglu, 2024). Moreover, the flexibility of OER allows educators to adjust content to fit specific pedagogical needs and contextual requirements – an invaluable aspect in STEM education, where curricula must stay current with scientific advancements. Most important of all, research indicates that integrating OER with active learning strategies significantly boosts student engagement and learning outcomes, as seen in the CREATEskills project, which reported improvements in students' problem-solving skills and motivation (Arabit-García et al., 2023).

However, despite these advantages, implementing OER in STEM education can bring challenges that need systematic attention. For example, many educators are either unaware of OER or lack experience in using them (Eaton et al., 2022). Professional development programs like workshops, funding opportunities, and recognition for OER authorship can effectively increase educator involvement with these resources (Hacisalihoglu, 2024). Additionally, concerns about the quality and reliability of resources persist even among educators who are familiar with OER (Yassin & Jung, 2021). Platforms like OER Commons (https://oercommons.org/) address these quality-assurance issues by offering curated collections of verified STEM resources suitable for various educational levels. Further, PLNs provide promising frameworks for the creation, adaptation, and distribution of OER, with informal learning within these networks supporting the validation, dissemination, and quality assurance of educational resources (Kleinschmit et al., 2023). Through these collaborative efforts, the educational community can maximize the potential of OER to transform STEM education while ensuring high standards for quality and pedagogical effectiveness.

Professional Learning Networks (PLNs) for STEM Teachers

PLNs are informal groups that enable and support the exchange of resources, knowledge, and ideas through interpersonal connections among educators. Within specific STEM PLNs, teachers actively assess, collect, and form shared knowledge on digital tools, all framed by personalization and social and digital connectedness (Oddone et al., 2019; Trust, 2017). These environments provide valuable opportunities for enhancing teaching practices beyond conventional onsite professional development (Peppler, 2017). Specifically, informal learning through PLNs has been found to improve teachers' STEM proficiency (Jackson & Mohr-Schroeder, 2018), encourage the use of new and effective teaching methods, and foster vital skills such as leadership, creativity, and social interaction (Terrazas-Marín, 2018).

STEM education is distinct from other subjects due to its rapidly evolving content and focus on student-centered inquiry, which promotes the adoption of technologies (Lesseig et al., 2016; Polgampala et al., 2017). Consequently, STEM teachers must retain both interdisciplinary content knowledge and the pedagogical skills necessary for integrative approaches highlighted in STEM education (Margot & Kettler, 2019). Additionally, teachers can find recommendations and strategies for planning and managing STEM activities, which often come with challenges, such as financial limitations and organizational systems (Aslam et al., 2018). With advancements in technology, STEM educators can utilize digital platforms, such as Facebook, to construct their unique PLNs and broaden their outreach beyond local communities (Carpenter et al., 2016). Furthermore, PLN-supported informal learning among STEM educators can enhance the quality assurance of OER. For example, Beaven (2018), in describing the OER lifecycle, emphasized that educators typically evaluate these resources before integrating them into their classrooms. As a result, PLNs may facilitate more effective use of and positive attitudes toward OER, highlighting the importance of teacher collaboration in selecting, utilizing, and adapting OER for successful implementation.

Types of OER for Online Professional Development of STEM Teachers

The exchange and evaluation of OER within a PLN can effectively meet the current demands of STEM teachers for educational materials, ensuring access to high-quality resources along with appropriate quality assurance (Almendro & Silveira, 2018). For example, teachers typically leverage OER to improve the quality of their teaching, particularly with regard to selecting and applying instructional strategies and guiding iterative instructional design (Jonassen et al., 1991). To that end, STEM educators have employed a range of OER materials, such as online textbooks, instructional videos, websites, and quizzes (Yassin & Jung, 2021).

Based on a synthesis of various types of OER identified in academic and professional references (Butcher & Moore, 2015; Hewlett Foundation, 2015; Jonassen et al., 1991; Pawlowski & Bick, 2012; Yassin & Jung, 2021), Park et al. (2025) found two major categories of OER that STEM teachers seek, share, and utilize within their PLNs: (a) *instructional materials*, including course content and lesson plans; and (b) *instructional resources*, which consist of strategies for instruction, professional development opportunities, and practical tips. (See Table 13.1.) In the following, we present a selection of resources within these categories that are tailored to support the needs of STEM education.

TABLE 13.1 Definitions of OER Types for STEM Teachers

Types	*Definition*
Course Material	Materials commonly used in classroom STEM instruction include videos, images, resources for interactive activities, worksheets, templates, etc.
Instructional Strategy	Verbal descriptions of intellectual resources connected to STEM teaching, drawn from experiences (lesson sequences, device usage manuals, etc.)
Lesson Plan	Lesson plans that can be used in instruction
Professional Development	Resources for STEM knowledge or skills
Tools and Tips	Details regarding instructional technology, open-source resources (like code.org), themes for special occasions (such as Halloween, Christmas, holidays, etc.), or descriptions and images of activities

Course Material

The category of course materials as part of OER offers direct support for classroom instruction, including instructional videos, images, hands-on activity resources, worksheets, and templates specifically designed for STEM subjects (Hilton, 2016). Characterized by their applicability to classroom settings, course materials enable teachers to integrate high-quality resources without the need for extensive modification. As the materials may be multimodal, they can facilitate the delivery of complex concepts. For example, visual resources, such as interactive simulations and diagrams, are particularly valuable in STEM fields where abstract concepts require concrete visualization (Wiley et al., 2018). Furthermore, the nature of multimodality enables accessible instruction, accommodating various learning abilities within STEM classrooms. The following are specific sources of course materials.

1 PhET Interactive Simulations (https://phet.colorado.edu/)

PhET, a website developed by the University of Colorado Boulder, offers free, interactive simulations for physics, chemistry, math, and biology. STEM teachers can utilize this space to visualize abstract concepts through engaging models.

2 OpenStax (https://openstax.org/)

OpenStax specializes in peer-reviewed and openly licensed textbooks for science and math courses for STEM subjects. The platform is particularly popular in secondary and higher education. All materials are allowed to be freely used, adapted, and distributed.

3 CK-12 (https://www.ck12.org/teacher/)

CK-12 offers a range of math resources from elementary principles to calculus and multiple science categories. It features customizable flexbooks for tailored lesson planning. A major helpful aspect of the website is that the content aligns with state standards.

4 National Science Digital Library (NSDL) (https://oercommons.org/hubs/NSDL)

This site aggregates resources from several major STEM sites, including NASA, Illustrative Mathematics (https://illustrativemathematics.org/), and TeachEngineering (https://www.teachengineering.org/). The materials are relatively high-quality across disciplines. STEM teachers can search materials by grade level and subject.

Instructional Strategy

Instructional strategies are cognitive tools that encapsulate educators' pedagogical expertise in STEM teaching. These verbal resources encompass lesson plans, device instruction manuals, and practical teaching methods developed from classroom experience (Tang & Bao, 2020). With an emphasis on procedural knowledge over content delivery, they connect theory with practice, helping new teachers adopt proven methods for teaching complex STEM concepts (Kimmons, 2015). These tools are especially useful for clarifying common misconceptions and overcoming learning challenges unique to STEM fields by offering practical, evidence-based solutions from actual classroom scenarios.

1 STEM Teaching Tools (https://stemteachingtools.org/)

The STEM Teaching Tools site, developed by the University of Washington, includes practical information for teaching STEM topics based on Next Generation Science Standards (NGSS, https://www.nextgenscience.org/) with research-backed approaches. In particular, it provides step-by-step instructional strategies, along with specific instructional strategies for various topics.

Lesson Plans

As categorized under OER, lesson plans serve as comprehensive guides for teaching STEM subjects by covering complex topics while preserving pedagogical quality. These resources typically contain learning objectives, materials lists, step-by-step procedures, assessment methods, and suggestions for differentiation (Blomgren, 2018). Their main advantage is their ready-to-use format with complete instructional packages that meet curriculum standards. Specifically, they offer a framework for incorporating inquiry-based learning, hands-on experiments, and project-based activities – key elements of effective STEM teaching (Hassler et al., 2018).

1 NSTA Lesson Plans (https://www.nsta.org/instructional-materials)

The National Science Teaching Association (NSTA) offers free lesson plans and book excerpts. Lesson plans are curated by following the NGSS and include assessments and classroom activities. STEM teachers can select the appropriate educational grades based on their teaching needs.

2 TRYEngineering (https://tryengineering.org/explore-resources/?resource_type=2054)

TRYEngineering is supported by *IEEE.org*. Materials are multimedia-enriched lesson sequences oriented toward cross-disciplinary connections. STEM teachers can search for lesson plans across all STEM subjects, organized by grade level, discipline, and topics.

Professional Development

Professional development resources as OER are tailored for STEM learning and skill-building. This includes online courses, workshop materials, webinar recordings, and self-paced modules focused on STEM pedagogy and content (Mishra, 2017). These OERs are characterized by their focus on both content mastery and pedagogical innovation and are of particular significance in STEM education due to the rapidly changing nature of STEM fields and teaching methods. As such, these resources support ongoing educator improvement while avoiding the financial and logistical barriers of traditional professional development (Allen & Seaman, 2014).

1 OER STEM Project (https://lincs.ed.gov/programs/oerstem)

This program, sponsored by the Literacy Information and Communication System (LINCS), provides online courses to train educators on STEM OER with a focus on the OER characteristics (i.e., Five Rs: retain, reuse, revise, remix, and redistribute). The resources are categorized by topic to help adapt materials to student needs.

2 National Science Foundation Resources (https://www.nsfresources.org/home.html)

The National Science Foundation Resources (NSF) website is somewhat outdated, but it still includes valuable resources and findings presented through research and development projects funded by the NSF. Primarily, the website addresses teacher professional learning resources and research-based STEM teaching methods.

Tools and Tips

Finally, tools and tips consist of practical OER that offer guidance on instructional technology, open-source platforms, thematic activities, and examples of real-world implementation. They include platform tutorials, such as those for Code.org, suggestions for special event activities, and documented classroom experiences with accompanying photos and descriptions (Santos-Hermosa et al., 2017). Tools and tips offer vital links connecting theoretical concepts with hands-on practice. They provide educators with proven strategies for integrating technology, which is essential for modern STEM teaching (Pitt, 2015). Moreover, they help instructors relate STEM principles to real-world situations and events, boosting student engagement and highlighting the relevance of STEM learning.

1 Code.org (https://code.org/en-US)

Code.org supports computer science (CS) education and artificial intelligence (AI) as a core part of K-12 education. The site offers free curricula with comprehensive teaching tools and specific guides for beginner-friendly coding and sequential learning. STEM teachers can implement this website directly to teach students computational thinking.

2 STEMfinity: Free STEM Resources (https://stemfinity.com/blogs/free-stem-resources)

While STEMfinity is similar to Teachers Pay Teachers (https://www.teacherspayteachers.com/), it focuses on STEM subjects with curated collections of learning materials. The site offers free STEM facts and resources, allowing STEM teachers to access tools and tips with minimal time and effort.

Technology Integration and OER in STEM Education

While the OER categories discussed above inherently involve digital technologies, it is worth examining how specific platforms explicitly support teachers in developing their TPCK. The intersection of OER and technology integration represents a critical area for STEM teacher development, as educators must not only access digital resources but also model effective technology use for their students. In other words, STEM teachers looking to build their technological knowledge and learn how to use technology in their teaching can benefit from online spaces by attaining OER, which combine high-quality digital content with practical training.

STEM teachers seek technological knowledge or technological PK through online PLNs to apply to their teaching (Margot & Kettler, 2019; Park et al., 2025). When teachers achieve OER for technology integration through online communities, it is possible to create customizable instructional resources and exchange best practices on the use of technology in STEM teaching among educators based on their educational needs.

Several OER platforms and projects are specifically designed to help STEM teachers enhance their technological knowledge and learn how to effectively use technology in their teaching. For example, Media Portal for STEM Teaching (https://medienportal.siemens-stiftung.org/en/media) offers more than 3,000 quality-controlled, freely accessible OER for STEM subjects, including digital teaching materials, practical examples, and instructional guides. Resources are designed to support technology-driven STEM lessons and are available in English and German. In addition, a specific webpage – Technology Integration in STEM Education (https://onlineprograms.education.uiowa.edu/blog/technology-integration-in-stem-education-tools-and-techniques) created by the University of Iowa – explores a variety of digital resources and educational technologies that support innovative STEM teaching, including practical examples and strategies for utilizing technology to foster engaging, student-centered STEM learning.

Conclusion

The potential of implementing OER in PLNs represents a paradigm shift in how STEM educators approach professional development in an era of rapid technological and scientific advancement. As demonstrated in this chapter, OER provide more than cost-free alternatives to traditional educational materials; they offer dynamic, adaptable resources that can be continuously updated to reflect current scientific knowledge and pedagogical best practices. The five categories of OER discussed here – course materials, instructional strategies, lesson plans, professional development resources, and tools/tips – collectively address the multifaceted needs of STEM educators across different career stages and teaching contexts. The informal professional learning ecosystem between OER and PLNs enables educators to transcend geographical and institutional boundaries, accessing expertise and resources that might otherwise remain unavailable. For STEM teachers who have entered the profession through alternative certification routes or who lack access to formal professional development opportunities, this combination is especially crucial.

Successful implementation of OER in STEM education faces persistent challenges, including awareness gaps, quality concerns, and the need for technological competence. However, the platforms and resources highlighted in this chapter demonstrate that solutions to these challenges are emerging through curated repositories, quality-assurance mechanisms, and user-friendly interfaces designed specifically for STEM educators. As educational institutions and policymakers recognize the transformative potential of OER and PLNs, increased support for these initiatives, such as integration of OER adoption into institutional strategic plans, professional development workshops and training programs on OER creation and implementation as well as providing practical and applicable technologies and resources, will be essential to ensure equitable access to high-quality professional development resources for all STEM educators.

> **KEY TAKEAWAYS**
>
> - **OER and PLNs can address critical STEM education challenges:** OERs provide flexible, cost-free solutions to the unique challenges STEM teachers face, including rapidly changing content, diverse student needs, and technological integration requirements. OER and PLNs provide crucial support for STEM teachers, bridging gaps between formal and informal professional development.
> - **Five OER categories provide practical guides in searching for instructional resources:** STEM educators benefit from five distinct types of OER: (a) course materials (videos, simulations, worksheets); (b) instructional strategies (pedagogical expertise); (c) lesson plans (comprehensive teaching guides); (d) professional development resources (online courses or workshops); and (e) tools/tips (practical implementation guidance). Specific platforms offer immediate, practical support for STEM educators, from interactive simulations to standards-aligned lesson plans and professional development modules.
> - **Future Directions:** Successful integration of OER in STEM education requires continued efforts to increase educator awareness, develop tailored resource collections, and strengthen connections between formal and informal professional development pathways.

References

Allen, I. E., & Seaman, J. (2014). *Opening the curriculum: Open educational resources in U.S. higher education*. Babson Survey Research Group.

Almendro, D., & Silveira, I. F. (2018). Quality assurance for open educational resources: The OER Trust framework. *International Journal of Learning, Teaching and Educational Research*, *17*(3), 1–14. https://doi.org/10.26803/ijlter.17.3.1

Arabit-García, J., Prendes-Espinosa, M. P., & Serrano, J. L. (2023). Open educational resources and active methodologies for teaching STEM in primary education. *Latin American Journal of Educational Technology – RELATEC*, *22*(1), 89–106. https://doi.org/10.17398/1695-288X.22.1.89

Aslam, F., Adefila, A., & Bagiya, Y. (2018). STEM outreach activities: An approach to teachers' professional development. *Journal of Education for Teaching*, *44*(1), 58–70. https://doi.org/10.1080/02607476.2018.1422618

Beaven, T. (2018). "Dark reuse": An empirical study of teachers' OER engagement. *Open Praxis*, *10*(4), 377–391. https://www.learntechlib.org/p/207177/

Blomgren, C. (2018). OER awareness and use: The affinity between higher education and K-12. *International Review of Research in Open and Distributed Learning*, *19*(2), 55–70. https://doi.org/10.19173/irrodl.v19i2.3431

Butcher, N., & Moore, A. (2015). *Understanding open educational resources*. Commonwealth of Learning (COL). https://doi.org/10.56059/11599/1013

Carpenter, J., Krutka, D. & Trust, T. (2016). PK-12 teachers' conceptualizations of professional learning networks. In G. Chamblee & L. Langub (Eds.), *Proceedings of Society for Information Technology & Teacher Education International Conference* (pp. 1936–1942). Association for the Advancement of Computing in Education (AACE). https://www.learntechlib.org/primary/p/171954/.

Eaton, C. D., Bonner, K., Cangialosi, K., Dewsbury, B., Diamond-Stanic, M., Douma, J., Smith, M., Taylor, R., Wojdak, J., & Wilfong, K. (2022). Sustainability and justice: Challenges and opportunities for an open STEM education. *CBE Life Sciences Education*, *21*(3). https://doi.org/10.1187/cbe.20-08-0180

Hacisalihoglu, G. (2024). *Open educational resources on STEM Center*. Florida A&M University. https://cst.famu.edu/departments-and-centers/OER.php

Hassler, B., Hennessy, S., & Hofmann, R. (2018). Sustaining and scaling pedagogic innovation in sub-Saharan Africa: Grounded insights for teacher professional development. *Journal of Learning for Development*, *5*(1), 58–78. https://doi.org/10.17863/CAM.26031

Hewlett Foundation. (2015). *Open educational resources. Advancing widespread adoption to improve instruction and learning*. https://hewlett.org/wp-content/uploads/2017/02/OER-strategy-memo.pdf

Hilton, J. (2016). Open educational resources and college textbook choices: A review of research on efficacy and perceptions. *Educational Technology Research and Development*, *64*(4), 573–590. https://doi.org/10.1007/s11423-016-9434-9

Jackson, C. D., & Mohr-Schroeder, M. J. (2018). Increasing STEM literacy via an informal learning environment. *Journal of STEM Teacher Education*, *53*(1), 4. https://ir.library.illinoisstate.edu/jste/vol53/iss1/4

Jonassen, D. H., Grabinger, R. S., & Harris, N.D.C. (1991). Analyzing and selecting instructional strategies and tactics. *Performance Improvement Quarterly*, *4*(2), 77–97. https://doi.org/10.1111/j.1937-8327.1991.tb00505.x

Kelley, T. R., & Knowles, J. G. (2016). A conceptual framework for integrated STEM education. *International Journal of STEM Education*, *3*(1), 1–11. https://doi.org/10.1186/s40594-016-0046-z

Kimmons, R. (2015). OER quality and adaptation in K-12: Comparing teacher evaluations of copyright-restricted, open, and open/adapted textbooks. *International Review of Research in Open and Distributed Learning*, *16*(5), 39–57. https://doi.org/10.19173/irrodl.v16i5.2341

Kleinschmit, A. J., Rosenwald, A., Ryder, E. F., Donovan, S., Murdoch, B., Grandgenett, N. F., Pauley, M., Triplett, E., Tapprich, W., & Morgan, W. (2023). Accelerating STEM education reform: Linked communities of practice promote creation of open educational resources and sustainable professional development. *International Journal of STEM Education*, *10*(16), 1–15. https://doi.org/10.1186/s40594-023-00405-y

Lesseig, K., Nelson, T. H., Slavit, D., & Seidel, R. A. (2016). Supporting middle school teachers' implementation of STEM design challenges. *School Science and Mathematics*, *116*(4), 177–188. https://doi.org/10.1111/ssm.12172

Lin, K. Y., Wu, Y. T., Hsu, Y. T., & Williams, P. J. (2021). Teachers' perception of STEM integration and education: A systematic literature review. *International Journal of STEM Education*, *8*(1), 1–19. https://doi.org/10.1186/s40594-018-0151-2

Liu, X., Chen, H., & Zhang, M. (2022). Trends and exemplary practices of STEM teacher professional development programs in K-12 contexts: A systematic review of empirical studies. *Computers & Education*, *188*, 104563. https://doi.org/10.1016/j.compedu.2022.104563

Ma, Y. (2024). Emerging trends and effective strategies in STEM teacher professional development: A systematic review. *Humanities and Social Sciences Communications*, *11*(1), 1–15. https://doi.org/10.1057/s41599-024-04272-y

Margot, K. C., & Kettler, T. (2019). Teachers' perception of STEM integration and education: A systematic literature review. *International Journal of STEM Education*, *6*(1), 1–16. https://doi.org/10.1186/s40594-018-0151-2

Mishra, S. (2017). Open educational resources: Removing barriers from within. *Distance Education*, *38*(3), 369–380. https://doi.org/10.1080/01587919.2017.1369350

Oddone, K., Hughes, H., & Lupton, M. (2019). Teachers as connected professionals: A model to support professional learning through personal learning networks. *The International Review of Research in Open and Distributed Learning*, *20*(3), 102–120. https://doi.org/10.19173/irrodl.v20i4.4082

Park, Y., Moon, J., & Na, H. (2025). Elementary STEM teachers' open educational resources and TPACK in a professional learning network: A case study. *Online Learning*, *29*(1), 192–213. https://doi.org/10.24059/olj.v29i1.4102

Pawlowski, J. M., & Bick, M. (2012). Open educational resources. *Business & Information System Engineering*, *4*(4), 209–212. https://doi.org/10.1007/s12599-012-0219-3

Peppler, K. (2017). Make-to-Learn: Broadening participation and deepening learning through making. In *Discussions on University Science Teaching: Proceedings of the Western Conference on Science Education* (Vol. 1, No. 1). https://ojs.lib.uwo.ca/index.php/wcsedust/article/view/4179

Pitt, R. (2015). Mainstreaming open textbooks: Educator perspectives on the impact of OpenStax college open textbooks. *International Review of Research in Open and Distributed Learning*, *16*(4), 133–155. https://doi.org/10.19173/irrodl.v16i4.2381

Polgampala, A.S.V., Shen, H., & Huang, F. (2017). STEM teacher education and professional development and training: Challenges and trends. *American Journal of Applied Psychology*, *6*(5), 93–97. https://doi.org/10.11648/j.ajap.20170605.12

Santos-Hermosa, G., Ferran-Ferrer, N., & Abadal, E. (2017). Repositories of open educational resources: An assessment of reuse and educational aspects. *International Review of Research in Open and Distributed Learning*, *18*(6), 84–120. https://doi.org/10.19173/irrodl.v18i5.3063

Tan, A. L., Teo, T. W., Choy, B. H., & Ong, Y. S. (2019). Understanding STEM teacher learning in an informal setting: A case study of a novice STEM teacher. *Research and Practice in Technology Enhanced Learning*, *14*(1), 1–18. https://doi.org/10.1186/s41039-019-0103-6

Tang, H., & Bao, Y. (2020). Social justice and K-12 teachers' effective use of OER: A cross-cultural comparison by nations. *Journal of Interactive Media in Education*, *2020*(1), 1–11. https://doi.org/10.5334/jime.576

Terrazas-Marín, R. A. (2018). Developing non-formal education competences as a complement of formal education for STEM lecturers. *Journal of Education for Teaching*, *44*(1), 118–123. https://doi.org/10.1080/02607476.2018.1422613

Trust, T. (2017). Using cultural historical activity theory to examine how teachers seek and share knowledge in a peer-to-peer professional development network. *Australasian Journal of Educational Technology*, *33*(1), 98–113. https://doi.org/10.14742/ajet.2593

U.S. Department of Education. (2024). *OER STEM project*. Adult Education and Literacy. https://lincs.ed.gov/programs/oerstem

Wiley, D., Bliss, T. J., & McEwen, M. (2018). Open educational resources: A review of the literature. In J. M. Spector, M. D. Merrill, J. Elen, & M. J. Bishop (Eds.), *Handbook of research on educational communications and technology* (4th ed., pp. 781–789). Springer.

Yassin, E., & Jung, I. (2021). Pedagogical purposes of using open educational resources in STEM education: A case study in a STEM high school in Egypt. *International Journal on Open and Distance E-Learning*, *7*(2), 99–116. https://ijodel.upou.edu.ph/index.php/ijodel/article/view/46

Zhang, L., Liu, K., Wang, B., Wang, Z., & Xu, Y. (2023). The effect of professional development on in-service STEM teachers' self-efficacy: A meta-analysis of experimental studies. *International Journal of STEM Education*, *10*(1), 1–25. https://doi.org/10.1186/s40594-023-00422-x

POSTSCRIPT

Woonhee Sung

As I write this postscript, I find myself reflecting on the conversations and collaborations that brought this volume together. The persistent gaps in science, technology, engineering, and mathematics (STEM) education, the rapid shift to online learning, and the ongoing quest to make professional development more meaningful for teachers – these are the challenges that drew us all to this work.

The book addresses online teacher education programs, training sessions, and continuing professional development (CPD) experiences, specifically guidance for the CPD instructors, faculty, and providers who design and facilitate professional development for STEM teachers. These online CPD facilitators face a unique challenge: They must not only demonstrate effective online teaching and facilitation strategies through their own online instruction, but simultaneously they must teach STEM-specific pedagogical methods – hands-on activities, project-based learning, laboratory experiences, and inquiry-based approaches – that are inherently interactive and experiential. That is, they face a dual challenge: How do you effectively model project-based learning in an online environment while teaching others to implement it in their own classrooms? How do you facilitate meaningful hands-on experiences through a screen?

Online CPD programs often adopt traditional approaches – uploading recorded lectures to learning management systems, assigning static readings, and providing individual or group assignments. But while these methods convey information, they fall short of the interactive, collaborative, and experiential learning that effective STEM teaching requires. That is, as demonstrated in the chapters in this volume, effective online CPD is far more than translating face-to-face experiences to digital formats. The models, strategies, and case examples presented here illustrate that successful online CPD requires intentional design that considers not only pedagogical principles but also the diverse contexts, needs, and identities of STEM teachers. With the application of diverse models and frameworks such as TPACK, ATTACK, and Design-Based Learning, each contribution reveals how thoughtful design can create transformative learning experiences for online STEM CPD.

What particularly resonates with me is how our contributing authors consistently foreground equity and inclusion throughout their work. The chapters on gender-equitable

DOI: 10.4324/9781003530343-16

teaching, multilingual approaches to STEM education, and place-based learning remind us that, on top of everything else, effective professional development must address the systemic barriers facing historically marginalized groups in STEM fields. These efforts represent more than best practices – they embody a commitment to justice and accessibility that must remain central to our work.

As we look toward the future, several themes emerge from this volume that warrant continued attention. First, the rapid advancement of artificial intelligence and emerging technologies will continue to reshape both STEM disciplines and the ways we prepare teachers to engage with these tools meaningfully and use them in the classroom. Second, the growing recognition of STEM education as inherently interdisciplinary will require new approaches to teacher preparation that transcend traditional subject-matter boundaries. Third, addressing the persistent inequities in STEM participation will demand continued innovation in our professional development approaches.

As we conclude this volume, I am reminded that the ultimate measure of our success is not found in the pages of this book, but in the classrooms where STEM teachers implement the ideas presented here and, ultimately, help develop a diverse and well-prepared STEM workforce. The journey toward more effective, equitable, and accessible STEM teacher professional development is ongoing. We hope this work provides both a roadmap for designing meaningful online CPD experiences and an invitation to join a community committed to transforming STEM education for all learners. We extend our deepest gratitude to all the authors who shared their expertise and experiences, to the countless STEM teachers who inspired this work, and to the readers who will carry these ideas forward into practice. The future of STEM education depends on our collective commitment to continuous learning, innovation, and equity.

INDEX

Note: **Bold** page numbers refer to tables and *italic* page numbers refer to figures.